# Design Science and Innovation

**Series Editor**

Amaresh Chakrabarti, Centre for Product Design and Manufacturing, Indian Institute of Science, Bangalore, India

The book series is intended to provide a platform for disseminating knowledge in all areas of design science and innovation, and is intended for all stakeholders in design and innovation, e.g. educators, researchers, practitioners, policy makers and students of design and innovation. With leading international experts as members of its editorial board, the series aims to disseminate knowledge that combines academic rigour and practical relevance in this area of crucial importance to the society.

Prabir Mukhopadhyay

# Ergonomics in Fashion Design

## A Laypersons' Approach

Springer

Prabir Mukhopadhyay
Faculty in Design Discipline
Indian Institute of Information Technology,
Design and Manufacturing
Jabalpur, Madhya Pradesh, India

ISSN 2509-5986          ISSN 2509-5994  (electronic)
Design Science and Innovation
ISBN 978-981-19-4536-6          ISBN 978-981-19-4534-2  (eBook)
https://doi.org/10.1007/978-981-19-4534-2

This Springer imprint is published by the registered company Springer Nature Singapore Pte Ltd.
The registered company address is: 152 Beach Road, #21-01/04 Gateway East, Singapore 189721, Singapore

*Late Mr. Dhirendra Nath Mukhopadhyay (Father),
miss you a lot!*

*and*

*Mrs. Meena Mukhopadhyay (Mother),
Dada, Mamoni, Ghantu and Leto*

# Preface

Ramp walking is very popular nowadays and is not confined alone to the female fraternity, but it has also percolated to the male fraternity as well. Fashion, lifestyle accessories, etc., are gaining all-time high popularity in India and all over the globe. The fashion industry is reaching new heights in terms of popularity and sales. The young generation is coming out of the "stylish" tag and becoming much more fashion conscious. The fashion industry is scaling new heights and foraying into serious domains like functional work wear design for the medical and defense sectors and special work wears for sports person, swimmers, wrestlers and many more. Jewelry which was a female bastion is now equally being embraced by the males and especially the younger generation. In the world of sports, shoes are becoming very important in terms of their design for increasing athletic performance. So, fashion or lifestyle accessory products which were once believed to be only for the rich and fashion conscious are now penetrating new serious domains of the society. Unfortunately, this serious domain is still not perceived that seriously like other branches of science, technology, or design. Though it is popular, but it still lacks that attention like other professions. This state of fashion and lifestyle industry reminds me of a story which I heard long ago when I was a kid. I narrate that story below.

There was once upon a time a man called Ram who lived in a village in remote rural area in India. He was famous for being "stylish" that he used to wear new clothes every day and flaunt it all through the village. The poor villagers used to look at him and discuss among themselves that how rich Ram was. One day Ram was riding a new bicycle wearing a special type of clothing called kurta (a very long shirt the length of which extends beyond the knees) and pajama (a modified form of trousers generally made of cotton and tied to the waist with a thread), when suddenly his kurta was caught in the wheels of the bicycle, and he fell on the ground. A truck was coming from the opposite direction at full speed, and before the truck driver could apply his brakes, the truck run over Ram and he was dead on the spot. The people around stopped the truck, took the driver out and handed him over to the local village head for justice. A court was summoned in the village to investigate whose fault was it that led to Ram's death. The truck driver tried to prove his innocence saying he had nothing to do as Ram came in front of the truck suddenly. The bicycle manufacturer was called, who also tried proving his innocence saying that his bicycle was structurally sound and there were no design flaws. He was questioned then how could the kurta get trapped in the wheels of the bicycle? He should have provided some cover over the wheels. To this, the bicycle manufacturer replied that it was a fault of the tailor who made that long kurta and not his. The tailor was summoned. He also tried to plead his innocence saying that he tailored the kurta for Ram not knowing that he would be riding a bicycle with such long kurtas. Everyone was confused as no one could pinpoint as to whose fault was it that led to Ram's death. This was because there were lot many factors involved in Ram's death, starting from the design of long kurtas to the manufacture of bicycle without any wheel cover to the bad road conditions to the driver of the truck driving at full speed. Due to lack of evidence, everyone was allowed to go. A primary school teacher was passing by at that time. The teacher yelled at the crowd and said everyone was at fault. The tailor should have known the context in which the kurta would be worn and designed it accordingly; the driver should have been driving with caution and be more vigil. Similarly, the bicycle manufacture should also have some idea about the different types of clothes people wear while riding a bicycle and design it accordingly. Everyone was stunned at this statement of the teacher. The fashion and lifestyle industry today stands at the brink of such a confusion and lacks a complete systems approach. If the fashion designers both budding and existing use some ergonomic principles in their design, they might be able to design their products which excel in design from the viewpoint of the users and the context of usage of the very product.

Fashion, lifestyle and accessory design are areas which are very creative, and people in these fields have a very bright future. As these are very user-centric designs, there is a need to understand the users, so that the design of such products is much more humane. Students and professionals coming and working in these areas do not always have a science or technology background. A good number of them come from arts and humanities. My students have been requesting me to write a book on ergonomics, with special emphasis on the fashion, lifestyle and accessory design industry. This book is upon their requests. This book is never a replacement for any other book on ergonomics available in the market that deals with the principles of ergonomics in depth. The purpose of this very book is to explain the principles of ergonomics pertinent to this industry in a simple, easy-to-understand and storytelling format, without the usage of technical jargons. This book can be used while students and professionals are designing or trying to ideate their products. To read this book, neither prior knowledge of the subject ergonomics, nor any knowledge of science and technology is needed. A few images and illustrations have been used. The illustrations are my own. The images have been picked up from royalty-free websites and are for representation purpose only and have been used to build the storyline. These images are not to be used for any design. This book can also act as a reference for those who are going for apprenticeship in these areas, the tailors, craftsperson related to this trade, etc. After reading the book, readers would get an idea about the subject ergonomics and how it could be applied in different areas of their profession.

Every chapter of the book starts with an overview, which prepares the readers for what they are expected to learn. At the end, the "ergonomic principles" discussed in the chapter are listed down. This is followed by assignments, and there are design directions for the same. There is a separate chapter dedicated to different types of exercises with directions for its solutions. The number of pages is also few compared to any standard textbook in the market. Thus, looking at these it should not make the budding designers and design professional in this industry to be jittery in reading. Rather it should encourage them to read it repeatedly. The small number of pages lets the user carry the book along while they are designing some products. As each chapter of the book is stand-alone chapters, the readers can jump on to any chapter of their relevance. But it is recommended that readers go through all the chapters at least once to get a good understanding of the application aspects of ergonomics in fashion, lifestyle and accessory design, as there is a dearth of books on ergonomics written in a simple language for professionals in these areas of design. Any criticism and suggestions from the readers are most welcome.

Jabalpur, India                                                    Prabir Mukhopadhyay M.Sc., Ph.D.

# Acknowledgements

Acknowledgement is due to many people who inspired me to write this book. My students have been inspiring me to write a book for fashion designers which are easy to understand without any technical knowledge of ergonomics. A very special thanks to Mr. Vipul Vinzuda, Faculty of Transportation and Automobile Design, Post Graduate Campus, National Institute of Design, Gandhinagar, India, for inspiring me to write this book and helping me in fine-tuning many of the figures. Acknowledgement is due to all the photographers from different websites for their kind permission to use their images for free. The line diagrams and illustrations have been created using google auto-draw online tool.

# Contents

# About the Author

**Prabir Mukhopadhyay M.Sc., Ph.D.** holds a B.Sc. Honours degree in Physiology and an M.Sc. degree in Physiology with specialisation in Ergonomics and Work Physiology from Calcutta University, India. He holds a Ph.D. in Industrial Ergonomics from the University of Limerick, Ireland. Prabir started working with noted ergonomist Prof. R. N. Sen at the Calcutta University both for his Master's thesis and later on a project sponsored by the Ministry of Environment and Forests, Government of India. He joined the National Institute of Design (NID), Ahmedabad, India as an ergonomist on a consultancy project for the Indian Railways. He later joined the same institute as a faculty in ergonomics. During his tenure at Ahmedabad, he worked on many consultancy projects. After working at NID Ahmedabad for two and half years Prabir left for Ireland for pursuing his Ph.D. in Industrial Ergonomics at the University of Limerick. After completing his Ph.D. he returned back to India and Joined the National Institute of Design, Post Graduate Campus at Gandhinagar and headed the Software and User Interface Design Discipline. After working there for around five years he joined his present Institute The Indian Institute of Information Technology Design and Manufacturing Jabalpur as a Faculty in Design Discipline. He still works there and is currently the Associate Professor and Head of the Discipline.

He is also the Principal Investigator of a National Council of Educational Research and Training (NCERT) New Delhi, funded project on ergonomic design intervention in schools. He has published around 23 peer reviewed journal publications, two book chapters and one single author book. He is a Member, Advisory Board, Centre for Creative Cognition, SR

University, Warangal, India. He has delivered expert lectures for many institutes like the Indian Institute of Technology Kanpur, National Institute of Design, Ahmedabad, Gandhinagar, Kurukhsetra and Amravati, National Institute of Fashion Technology, Jodhpur, Delhi Technological University, etc. email: prabir@iiitdmj.ac.in.

# List of Figures

# Introduction to Fashion Ergonomics

**1**

**Overview**
This chapter introduces the readers to the complex world of ergonomics and its relationship with fashion. The components and application areas of ergonomics related to fashion are outlined here. After going through this chapter, the readers would get an overview of the application of ergonomics in fashion and lifestyle accessory design. This chapter also gives an overview of the systems perspective of ergonomics and how it can be applied to the fashion industry. The different component which comprises a system and how each influences the apparel industry is discussed in this chapter.

Clothes are believed to be the second skin of man, and apart from just covering the human body it protects the body from the natural hazards of the environment. If one looks at the genesis of clothing, its main purpose was protection. So, clothing to the accessories that people wear is like a government in a country of the people, for the people, and by the people. This is what we call people-centric fashion and lifestyle design. Unfortunately, in the fashion industry there is very little emphasis on the "user" but more on the other aspects of it. It is here that the discipline ergonomics can come at the rescue. Ergonomics in a layman language means the interaction or the relationship between man, product, and the environment, all with an aim to ensure comfort and well-being of the man. It is a holistic approach to investigate the well-being of man by trying to relate the different aspects around man and how it influences his actions, behavior, thought process, and overall well-being. The theory of wearable attitude indicates few major principles for the second layer. It talks about modesty, which reflects the background and "taste" of the users. The third principle is adornment which in essence was the sense of belongingness of the attire or work wear with the user. The fourth principle is the hierarchy principle. This was the flow of the gradual development or comprises theory to understand the human nature to cover their body and feel secure. More and more the society was becoming larger and more complex from various angles, these all effected directly or indirectly to the society in micro and macro-level. For example, if we observe the decoration of body with flower was replaced by the natural objects like pearl, small conch shell, bone horn, etc., and much later, the metal was used to take form of ornaments because the human mind was looking for some idea of permanent solution of accessory in terms of adornment and hierarchy. The invention of more new materials brings the scientific approach toward our nature. We do not appreciate to wear heavy drape clothes and heavy metal jewelry in present time, because the lightweight dress materials, jewelry, wristwatch, shoes, etc., all are available in the market. This is possible by a synchronous fusion of science, art, and technology. The artistic part is known as "fashion", and the amalgamation of science, art, and technology making it user centric is known as "ergonomics".

When clothes and accessories are designed, there are lots of "human" elements involved which are times ignored by many. These elements are the dimensions of the human body (called anthropometry) which are immensely important. Next element is the physiological property of the human skin which would dictate the type of fabric to be used in a specific climate. The next important element is biomechanics of the human body, i.e., the movement of the body and its different parts which should decide the allowance in the clothing and the strength of the fabric and the stitches in those parts which would be susceptible to maximum force and pressure by the bony joints and the tissues. So, in essence there is man at the center, we then have apparel, there is environment, and each is related to one another. If one must do a good design, then each aspect of the human body needs to be considered in detail (Fig. 1.1).

Each wearable material has its own characteristics and that is manipulated or exaggerated to enhance our look all together. Style and fashion both are important factors to consider along with functionality aspect and feel-good factor. From head to toe we have created lots of wearable components to protect our body in daily life and in workspace. The consumers do not know that the scientific research takes place behind the product, but they remember them because they are comfortable with it and feel confident. So, can we question ourself what are the parameters of liking and disliking of aesthetic phenomenon

**Fig. 1.1.** Role of ergonomics in fashion

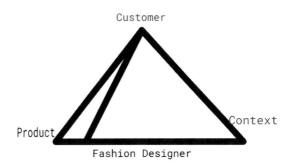

in humans? And the second question is how can we resolve them through fashion and ergonomics?

## 1.1 Ergonomics and Fashion

In the realm of fashion industry, ergonomics would be the interplay between apparel, man, and the environment. There are many different levels at which ergonomics can be applied. At the first level it is applicable when the apparels are being designed. One must consider the different human body dimensions, the biomechanics, and physiology of the human body. The second level is at the manufacturing sector where people are working. To enhance productivity, one can apply ergonomic principles like proper workstation height, illumination level, duration of a shift, work–rest cycle, all with an aim to increase productivity and minimize errors and rejection in the manufacturing process. The third level of intervention is at the level of displaying of fashion elements in the showrooms as well as in open space, because in India we are accustomed with roadside market which is more active than any other market or showroom. It can be a standardized display system which will serve the local vendor and shopkeeper to maintain a quality of display system. Ergonomics will dictate the exact angle at which it is to be displayed, the height, illumination level, proper labels, etc., so that it is able to attract a person entering a showroom or moving in the market. Thus, ergonomics can be applied in the fashion industry at each level to make the product much more humane Fig. 1.2.

## 1.2 Components

There are many different components that make up ergonomics related to the fashion industry. The first is anthropometry, which deals with human body dimensions. The second component is biomechanics, which deals with movement of different body parts. The third component is physiology, which deals with the structure and function of

**Fig. 1.2** Ergonomic interventions in fashion

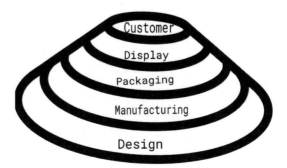

different parts of the human body. The third important component is psychology which in essence deals with behavioral issues of humans. The fourth component is technology, and the fifth one is mathematics/statistics, which helps in validating data collected from the field before it is handed over to the manufacturing unit for mass scale manufacturing.

## 1.3  Application Areas

The application of ergonomics in this domain is just enormous. From developing "size charts" for the population to designing of individual work wear and accessories, ergonomics can be applied everywhere. Ergonomics will dictate the type of stitching at the joints of the fabrics which would facilitate movement and at the same time be robust. Ergonomic principles can be applied in packaging and displaying of merchandize so that they are able to draw the attention of the customers. In manufacturing ergonomic principles would dictate the height of the workstation, illumination level, and work–rest cycle duration to improve quality of work and productivity.

## 1.4  Ergonomics as a "Grandmother" in the Family

The role of grandmother in a joint family is to keep the entire family united, and it is possible by keeping a tab on the well-being and characteristics of each family member. Ergonomics also takes just a holistic approach in the "fashion family" by making the design more "user centric". Fashion is for the user, so ergonomics will guide and dictate the type of shape, color, texture, and material based on the characteristics, aspiration, and limitations of users. Ergonomics can tell us what can and will go wrong by looking at the entire discipline of fashion from the user's point of view. To do this ergonomics views fashion in three different components. First is the product, followed by space or the context in which the product is used. The third is the most important part which is interaction, the interplay between user and the fashion product.

## 1.5  Systems Ergonomics in Fashion Design Lifecycle

Anything having more than one entity is called systems. The interplay between the different components (subsystems) of the system is vital for the wearable design world, and it is at this juncture that ergonomics plays a very big role. For example, a wearable system is composed of style elements (inspiration, silhouette, color, fabric, trims, mechanism of opening and closure of the garments, functional aspect for the target audience), the context or space in which these are displayed (visual merchandizing), the manufacturing process, packaging, marketing, service, and personal grooming of the salesperson. So, if the apparel industry must benefit in the form of a facelift from the viewpoint of ergonomics, then all these different components must be addressed in minute detail. In other word this macro-view or zooming out and zooming in feature of the entire fashion industry must be perceived. Zooming in is when one looks at different components of the system, and zooming out is when one looks at the individual components as a part or in group as one and unified whole. Ergonomics guides the fashion designer to investigate fashion design from this novel perspective at every stage of the design process. This type of approach ensures that ergonomic issues are factored in the design at a relatively early stage rather than noticing late.

## 1.5.1   Systems Perspective of Ergonomics

In fashion design a holistic perspective of the entire fashion design from data collection, analysis, translating the same into design solution, and manufacturing for the target users is essential. This ensures that the designer has a holistic view of the entire design life cycle with specific emphasis of the human intervention at different points.

## 1.5.2   Parts of Systems Ergonomics in Apparel Industry

Normally systems ergonomics comprises the following parts, and they are as follows:

Work environment.
Physical environment.
Technological environment.
Psychosocial environment.

The work environment in this case could be the manufacturing unit where the garments are manufactured, or could be the showroom, or open markets where fashion merchandize is displayed. Hence the allied elements of the work environment could be the machineries, the physical environment like heat, light, and ventilation inside the shop floor. The other components of the work environment could be the elements of dust, smoke, and fumes in the environment. The work environment also involves the organizational structure, hierarchy if any, power of the supervisors and subordinates, and the relationship among the workers and related matters. It also deals with employee benefits, medical reimbursements, leaves, and compensation to the staffs.

Physical environment mainly focuses on the physical aspects of the environment like heat, light, humidity, ventilation, etc., which influences the manufacturing process. These aspects of the environment also affect the look and feel of a fashion product and inside a showroom might play a significant role in influencing the customer to select or not to select a particular product. The technological environment is that which comprises machineries on the shop floor and the different gadgets in the showroom or open markets like lights, speakers (for music), public address system, etc. Psychosocial environment comprises the user's behavioral pattern both inside the shop floor and in the showrooms and open market. This plays a crucial role in productivity and in buying pattern of the users.

The principles of systems ergonomics can be similarly applied in different apparel industrial sectors. In the knitwear garment sector this can be applied very effectively. Placement of the machines and the relative distance to be maintained between them are related to quality and productivity of the task in question. Next comes the exact height at which the task should take place relative to the human body and then comes the issue of amount of force requirement if it is a manual machine and the degree of precision required in case of an automatic or semiautomatic machine. The environment in terms of heat and humidity affects worker productivity and hence needs to be considered not only for productivity but also for quality of the task as before. The last systems ergonomic issue is illumination level which needs to be considered in tandem with the age of the workers and the movement pattern of the workers in between darker and lighter areas of the shop floor if any.

In whatever industry we are talking about whether it be the Denim manufacturing, medical wear manufacturing or kids wear manufacturing units, we need to focus on a few ergonomic attributes:

1. First decide the work surface height. To do this take a call as to the type of task being performed, precision, light or heavy.
2. Once you decide upon the type of task fix it as per the relevant landmark in the body, for example, if its precision fixes it near the sternal notch (the depression in the breastbone), if light work fixes it near the elbow, and if heavy fixes the work surface near the trochanter or the place where you wear your trouser (hip region)
3. After deciding the height, you need to decide upon the depth and the length of the work surface.
4. The depth of the work surface is decided by the forward reach while seating or standing.
5. The length of the work surface can be decided by taking the span akimbo (spreading the two hands on either side and the distance between the tip of the fingers of the left and right hands).

## 1.5.3   Application of Systems Ergonomics

It is seeing that if a man is not comfortable in wearing a black-colored kurta does not only mean that the tailoring is bad. It could be also since black is considered "evil" in many communities in India. The current scenario of fashion industry dwells on this fundamental principle, by ignoring the ergonomic issues in the overall systems such as user, environment, and the interplay between user, environment, and the machine which is used to manufacture.

**Fig. 1.3** Systems perspective
in fashion ergonomics

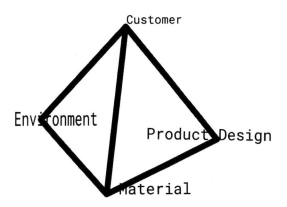

In the manufacturing sector this is evident in terms of increasing productivity. In the marketing sector this concept has great significance. In the showrooms and in the open market for fashion accessories, systems level approach can boost sales. It is not only important to keep fashion products, but more so how to display them in an effective and efficient manner keeping in mind the illumination, background elements, comfort level of the environment in case of a controlled environment. In case of an open market, ergonomic principles like position of the sun, wind direction, etc., can go a long way in creating comfortable environment, which would lure the customer. Adding to this is the personal grooming (also called personal ambience) of the staff members selling the products. This is the overall "systems ergonomic" approach in selling these products. The advantage of such an approach is that it gives an insight into the type and quantum of complexity in the system and hence paves the way toward its rectification (Fig. 1.3).

## 1.5.4    Characteristics and Pertinence

System has a purpose, and this is the main aim of the entire apparel industry if we see it from this perspective. The main purpose is business with quality product so that customers are satisfied, and it builds a good brand image of the company. The second characteristic is hierarchy. This is applicable at the organizational level where superior and subordinates play a very important role, especially in terms of their relationship and decision-making capability. The third characteristic is that all systems work in a particular context or environment. The manufacturing sector has a physical environment in which the work is done, and similarly the open market and showrooms also have an environment in which all activities take place. The fourth characteristic is that each component of the system serves specific function. In a showroom or open market system, the sales staff, the illumination, and the displays each have their specific roles to play. This brings to the fifth characteristic which is the interaction between different components for a common purpose. If we take the showroom, then every element of the showroom like staffs, customers, and the environment inside all interacts with one another in some way or other. The interaction of the ambience and the staff behavior gives a unique experience to the customers. You go closer to a showroom or an open market when something different falls within your visual cone. You move closer and then either enter the showroom or pick up the product in an open market. The salesman greets you with a sweet smile and is neatly dressed; you are at ease and feel homely. So, if these elements were not in place, let us imagine bad displays, insufficient illumination, and badly dressed salesperson and ill mannered, then you no longer feel comfortable and move away. The company loses a valuable customer.

## 1.5.5   The Pyramidal Structure of Systems Ergonomics

The structure of systems in the context of apparel/fashion could be imagined to be like a pyramid. The three corners would represent man, product, and environment. It has an apex and a base. At the apex or the sharp end of the pyramid lies the users (customers) for whom we design. At the base of the pyramid (also called the blunt end) lies the designer. So, for overall success of the system, it is necessary for designers to come out of the blunt end of the system, put themselves in the feet of the users, and design their products accordingly. This is achieved through user study and through the application of ergonomic principles. Wardrobe malfunction is a well-known incident in the fashion industry and happened with many eminent personalities. It is not only embarrassing but at times hurts the sentiment of others and can lead to bigger disasters in the society. The reason for such wardrobe malfunction is not that the user could not wear the garment properly, but more so because the designer who designed the garment was at the blunt end of the system and was least bothered to come at the sharp end to put himself/herself at the feet of the user. Thus, it was an example of total systems collapse and not just an error on part of the user. Since users and designers always remain at two different ends of the system, this gap needs to be bridged. Ergonomics in fashion helps to bridge that gap and thus makes the design much more humane. Let us take the example of "Miss India". Many a times after being crowned we have seen the crown falling from the head when the participant turns back, or some participants must hold the crown with one hand. This is because the designer is at fault by not considering the context in which the crown would be used. In such functions lot of dynamic body movements happen. These should be factored in the design of the crown so that it holds to the head amidst all such movements like bowing, hugging, etc. So, the sharp end of the pyramid here is the participant (Miss India) and at the blunt end of the system is the designer who sits in his/her air-conditioned office and designs the crown, without thinking of the context of use.

## 1.6  Ergonomic Principles

(a) Fashion or lifestyle products cater to human desirability, whereas ergonomics relates to its functionality from the user perspective.
(b) Ergonomics is that lens which gives a much broader view of fashion going much beyond the product and moving into its context of usage.
(c) Ergonomics helps to attain fashion another level by giving it that "humane touch".
(d) Fashion design must be looked from a systems perspective which is the macro-picture in which it is positioned
(e) The parts of the system that is the subsystems are as important as the system as a whole.
(f) This systematic approach helps in better understanding of the ergonomic problems in terms of its nature, location, and quantum.
(g) In fashion it is important to look into the design productivity but keeping an eye on the user who is an integral part and embedded in the system.

## 1.7  Assignments

(a) Pick up any garment of your choice and identify the human touch points in them. List down the good and bad ergonomic issues in them. Select an article from your wardrobe what you wear quite often; bring out the attire which you hardly wear. Do a comparative study and find out the reason behind this liking and disliking and put them on paper. We shall categorize the issues into parts—design (part one) and ergonomics (part two).
(b) Select an accessory product like lady's side bag, wristwatch sunglass, etc., from roadside market, and branded stuff and identify the elements of product, space and interaction, and the role played by ergonomics at each stage, if at all it has been considered or implemented, does it really enhance the style, design, and function for better utility purpose.
(c) Now learn the scientific approach to analyze the loopholes by the principle of wearable ergonomics (or similar something like that) to identify the faults in the product and find out the solutions on spot. For example, from a "grandmothers" perspective analyze the hat worn by Sherlock Holmes focusing on the different ergonomic issues and styles related to fashion, including visual merchandising, design, and manufacturing.
(d) Select a showroom or small unit of apparel production unit and analyze it from a systems ergonomics perspective, identifying the sharp and blunt end.

(e) Put all the observation on a chart and tally with the standardized system of system ergonomics and if not, what could be the creative and scientific solution to it?

(f) Select an open market area where fashion items are sold and try to identify what are the vulnerable areas which can lead to the entire systems collapse.

(g) Next step is to strengthen the vulnerable areas more effectively to maintain the system design.

(h) Now we can compare in Indian context what could be the basic understanding of the visual ergonomics for showroom as well as street vendor for visual merchandizing in India and how to implement them to make it cost effective.

## Bibliography

Cho DJ, Han KS, Hwang KH, Chung KY, Lee JH (2007) Apparel coordination based on human sensibility ergonomics using preference of female students. In: Proceedings of the Korea contents association conference. The Korea Contents Association, pp 146–150

Elsayed WA, Eladwi MM, Ashour NS, Shaker RN, Shaheen E (2019) Ergonomics approach for fashionable apparel design. Int Design J 9(3):273–280

Erdinc O, Vayvay O (2008) Ergonomics interventions improve quality in manufacturing: a case study. Int J Ind Syst Eng 3(6):727–745

Mukhopadhyay P (2019) Ergonomics for the Layman: applications in design. CRC Press

# Human Body Dimensions in Fashion and Lifestyle Accessory Products

**2**

**Overview**

This chapter deals with application of human body dimension in fashion. The different aspects of body dimensions and its pertinence to different types of garment design are discussed in detail. This chapter also gives an insight into different ergonomic issues that goes into fashion wear for the different body parts and hence touches upon summer and winter wear as well. Anthropometric, physiological, and other structural issues of the human body are discussed here.

## 2.1 The Need for Dimensions

We live in a world where every individual is different. This difference exists in physical form such as dimensions of different parts of the body. That is why some people are tall, some short, some are thin, and some are fat. Some have more strength and some have less. So, it is a complex world, full of variety of people. For fashion industry to survive the biggest challenge is how to design product or products that fits all with ease and comfort. It is always not possible to design for each individual person in this world. There are garments which must fit all. There are shoes which must be comfortable for all. So, the option in front of the fashion designer is to generalize the human body dimensions and come up with certain standard design that fits the entire spectrum of the population. This is where anthropometry or the measurement of human body dimensions comes into the picture.

© The Author(s), under exclusive license to Springer Nature Singapore Pte Ltd. 2023    11
P. Mukhopadhyay, *Ergonomics in Fashion Design*, Design Science and Innovation,
https://doi.org/10.1007/978-981-19-4534-2_2

## 2.2   Introducing Anthropometry

This is the scientific measurement of the different parts of the human body with the aid of specific scientific instruments. The specificity of such measurement is these are taken with reference to certain landmarks in the human body. The landmarks are essentially bony elevations or depressions in the body. In anthropometric dimensions, the measurements are taken either from bony elevation to elevation, elevation to depression, or from depression to depression. You have noticed how difficult it is to walk with a shoe which is bigger or smaller than your feet. A suit which is bigger than your body not only makes you look a little funky, but unnecessarily draws the attention of others, and you lose self-confidence too. Thus, scientific measurement of different parts of the human body plays a role in fashion and accessory design (Fig. 2.1).

**Fig. 2.1** Users in odd fit with garments. *Source* Photo by SHVETS production from Pexels https://www.pexels. com/photo/cheerful-slender-woman-in-oversized-pants-6975488/

## 2.3  Types of Body Dimensions

There are two types of dimensions, namely static and dynamic. As the name suggests, static dimensions are taken with the body in the resting position. Dynamic dimensions are taken considering the different movements exhibited by the body. So, in essence all dimensions are initially static but move into dynamicity as the body/body parts start moving. If your shirt sleeves are designed with the hands hanging by your side, only then there could be a problem when you raise your hand above your head (as you do when traveling in a crowded bus). In that case your wrist and forearm might be exposed leading to embarrassment in public domain. This is an example where dynamic anthropometry can play a big role. While taking human dimensions, one must take the static dimension and give adequate allowance for movement that the body will exhibit. If you take out your leather belt after wearing it for some time, you will notice that the belt is no longer straight by a little curved at the waist. This is because of the dynamic movement of the waist (Fig. 2.2).

**Fig. 2.2** Different type of anthropometric dimensions in fashion. *Source* Photo by SHVETS production from Pexels https://www.pexels.com/photo/fit-woman-measuring-waist-with-tape-6975475/

## 2.4   Measurements

There are different techniques for measuring the different human body dimensions. Traditional methods include the usage of Martin's anthropometric set, sliding and spreading calipers, and measuring tape. These are all direct methods for measurement, where the investigator touches the human body. There are indirect methods where the subject stands against a grid board, and the number of grids is counted to know the exact dimensions. Lately 3D scanners have arrived which again is an indirect method of measurement, and it is very accurate and faster than all other methods (Fig. 2.3).

## 2.5   Percentile Concept

If you tell someone that you have got 70 marks out of hundred in mathematics in your class, it might first sound to be impressive. If the person is a teacher, then he/she might ask you another question that what is the highest marks in the class and the lowest marks as

**Fig. 2.3**  Measurement techniques in anthropometry. *Sources* **a** Photo by SHVETS production from Pexels https://www.pexels.com/photo/sporty-women-with-different-bodies-covered-with-measuring-tapes-6975547/; **b** Photo by Monstera from Pexels https://www.pexels.com/photo/cheerful-multiracial-girls-measuring-each-other-5063385/

**Fig. 2.4** Percentile concept in fashion design

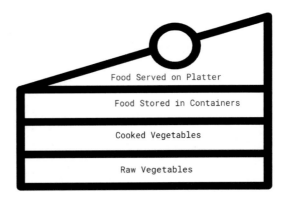

Food Served on Platter

Food Stored in Containers

Cooked Vegetables

Raw Vegetables

well. The reason being that 70 marks might be the lowest marks in the class of 100 students. So that is not at all good performance. If you would have got 55 marks when the highest marks in the class were 58, then it really speaks well about your performance in the class. Thus, what transpires from here is that just marks in the class do not reflect your performance, it must be seen against the marks scored by other students in the class as well. What is important is where you stand among all the 100 students, at what position. This position is known as the percentile concept. Percentile means out of 100. Anthropometric dimensions are always available in the form of percentile data for ease of application in design and more so in fashion design. For students who still do not understand the percentile concept I give them this example and analogy. You are hungry and your mom asks you to get some vegetables (raw) from the market. This raw vegetable is your population for whom you are to design your garments. The vegetables need to be cooked. Your mother adds spice, oil, etc., to cook. This spice and oil are the statistical treatment of the data from the population. Still this food is not ready to be taken unless it is taken out of the utensil and served on a platter. The statistically treated data is not usable unless it is "served" properly. Your mother now serves you the cooked vegetable on a platter along with other food, which can now be taken in quantity you like. This is when statistically treated data are arranged on a scale of 0–100 for the ease of your use in designing a product (Fig. 2.4).

## 2.6 When to Opt for Which Percentile?

The general rule of the thumb in percentile value selection in anthropometry is that when it comes to designing something which has to do with access or reach, we opt for lower percentile values, i.e., values below the 50th percentile. When it has to do with clearance, we opt for higher percentile values. If you are to design only one trouser (crazy though!) for the thin and the fat person, then what would be your approach? From ergonomics perspective you will select a value for waist circumference toward the higher side. The reason being your first aim should be the fat person, if he is able to wear it then the thin

**Fig. 2.5** Percentile value selection. *Source* Photo by Aenic Visuals from Pexels https://www.pexels.com/photo/woman-carrying-baby-on-her-back-3702508/

person will have no problem in wearing (though will be a little uncomfortable). So, when it comes to clearance like in this example we opt or take off from the higher percentile value. If you are to design a necklace for a tall and a short person, then you need to consider till what point in the body you would allow the necklace end to dangle. If it is too long, it might touch the naval and look odd. On the other hand, a little short necklace would not look that bad (we are imagining the situation as of now!). In this case the height of the necklace must be considered by considering the stature of the lower percentile (shorter) person. So, when it comes to access or reachability, one must opt for the lower percentile value (Fig. 2.5).

## 2.7  The Myth About the 50th Percentile

I have often experienced that design students after going through all these concepts try to play safe with the 50th percentile anthropometric values. They think that if they take the middle value for all anthropometric dimensions, then they will be able to cater to the need of maximum people. This type of thinking is wrong. The human body dimensions are not in absolute proportion for anyone. This means that the body is a mixture of different percentile values for different parts for the same individual. My stature could be 95th percentile, waist 50th percentile, feet length 75th percentile, and upper arm circumference 5th percentile. So, if you go by the uniform you might end up with the wrong dimension.

## 2.8  Using the Percentile Data

Fortunately for the designers there are anthropometric database already available, and we can reduce the tedious task of collecting data to converting them into percentile values. There are normally certain steps involved in calculating the percentile value for a given garment/wear. These are as follows:

Step 1: Let us take the example of a bracelet for the upper arm. You must start doing a "task analysis", i.e., observe how many ways people wear a bracelet and under what different contexts. You might notice in this step that people put on the product in different ways, and with different parts of the palm, and with different degree of force. This stage helps you to identify the relevant anthropometric dimensions in the design; in this case it is the bracelet.

Step 2: You now identify different touch points, draw them, and study them very closely. Then you try to establish a link between these touch points in the body and the closest possible anthropometric dimension in the database. For example, if you are to list down the relevant anthropometric dimensions for designing a bracelet, then the probable anthropometric dimensions relevant to its design could be upper arm circumference.

Step 3: Once the anthropometric dimensions are identified, go to the relevant table of data for each dimension. A close look at the table of data will reveal that there are five different percentile values given for male and female population in the Indian context. For other parts of the world, it might be different. So, there are male values, female values, and combined values (combination of male and female values). Write down the table of data for each dimension.

Step 4: Ask yourself who is your users? Male, female, or both male and female? If it is only male, things are relatively simple than if it is both male and female users. So let us start with the complex case first and imagine that our target users are both male and female. So, you are now ready with the relevant anthropometric dimension of each body part and their percentile values. Now is the issue of value selection and optimization.

Step 5: You learnt the two-thumb rule in percentile value selection, for access lower percentile, for clearance higher percentile value. These are the "takeoff" points for percentile calculation, and we will never fix the dimension at this. For the 5th percentile value for both male and female look for the lowest value. If female value is lower circle around it, this becomes your lower percentile takeoff value. Again, for the maximum or 95th percentile (in case of Indian population) select the higher value among the male and female 95th percentile. Circle around the male value if it is higher. Now you have a range of percentile value, female lowest to male highest. You must now optimize your dimension within this range.

Step 6: Now look at the intermediate value for both males and females. If it is a question of access, try to optimize it a little toward the lower percentile, and in case of clearance try optimizing it a little toward the higher percentile value. Do not forget to add allowance for dynamicity of movement, if it is to be worn above clothing, expansion, and contraction of muscles when the hand moves and so on. This will depend on your user study and task analysis done in step 1. Remember to fix your dimension to the nearest whole number for manufacturing feasibility. So, if your anthropometric dimension turns out to be 112.2 mm, then round it off to either 110 mm or 120 mm depending upon your observation and task analysis.

## 2.9  Anthropometric Dimension; Some Caution

You must remember that the data in the book is a static data. You must make the data dynamic by adding dynamicity (Fig. 2.6) into it. This is achievable only through proper task analysis with the user and the product. For example, if you are to design a jacket for a college teacher, then to decide upon the sleeve length of the jacket it is not enough to only tally the dimensions of the upper and forearm. These dimensions would be only the static one. The teacher while teaching would write on the board and the hand would go up. So, in this process the sleeves should not roll back to a significant extent to look odd. This dynamic movement of the hand must be factored in your dimensions over and above the static value. This is achievable only after careful task analysis of the user with the garment. So, while using anthropometric dimensions from the database and using it in design, one must be careful in their application. Data from the database need to be optimized from much different perspective, after a careful and thorough task analysis.

The human body can be categorized based on the amount of muscle, fat, and bony material present. This type of classification is called somatotyping. There are three major types of body somatotypes: ectomorph, mesomorph, and endomorph. Ectomorph is a very skinny person. Endomorph is a bulky person with lot of fat deposit in the body. Mesomorph is a muscular person. These somatotypes need to be considered while designing garments and adorations including functional work wears. In the subsequent chapters we are going to discuss anthropometry in greater detail in specific context of different types of design. Other issues like physiology, psychology, etc., will also be dealt with.

**Fig. 2.6** Dynamic
anthropometric dimensions in
fashion. *Source* Photo by
Wellington Cunha from Pexels
https://www.pexels.com/photo/
woman-standing-while-
wearing-blue-sleeveless-dress-
1918445/

## 2.10   Ergonomics of Fashion Wear for Different Body Parts

Ergonomic principles can be applied for designing fashion wear for different parts of the body. This would ensure that these are comfortable, easy to wear and take off and fit the body parts of the users and function as an integral part of different body parts for which it has been designed.

## 2.11   Adorations for the Head

1. *Thermoregulation*: The human head is composed of the outer shell called the skull which protects the vital element called brain. It is strategically located on the topmost point on the body and is the main center for coordination of the body. The skull is covered by the skin and on top of that we have hair which adds beauty to the head. It is from the head that about 30% of the body heat is lost. So, any design for the head

should respect this. Especially in a tropical country where the environment is hot and humid heat loss is very important as cutting of direct solar radiations to the head. So, hats need to be designed in such a way that there is passage for heat escape as well as avenues for shielding against solar radiation and other hazards. There is maximum concentration of sweat glands here which acts as safety valves to emit heat from the body.

2. *Anthropometric dimension*: As mentioned in the previous chapter a good fit is essential for the adoration to stay in place securely. Depending on the type of task to be performed the relevant anthropometric dimensions of the head plus allowance for hair must be given. The forehead skin exhibits lot of dynamicity, when a person communicates, reacts, or speaks and that moves the skin of the head as well. So, you must account for this movement. People wearing adorations in social gathering exhibit lot of movement and many a times the adoration falls. Proper dynamic data would prevent that.

3. *Comfort factor of the head*: The skin of the head including the forehead is very sensitive, and hence the inner lining of the adorations needs to be well padded for additional comfort. You might have noticed that emotional attributes in the society are manifested by kissing on the forehead and elders touching juniors head for blessing them. This is because any mild touch stimulus in this zone elicits sensation all over the body.

4. *Structure of the head*: There are wide variations in the structure of the head (skull to be specific), and this must be accounted for in design. The head is not a perfect circle, and majority of the hat /head gears found in the market seem to follow this wrong notion in their design. The surface of the head is convex that too the degree of convexity is not uniform for all individuals. The skull is a little broader at the back and tapes a little at the front. This must be accounted for while designing your product. The shape of the head adoration should fit the different geometrical shape of the head and that would add to the first level of comfort. There are differences in the structure between male and female skull. For example, it has been found that males have bigger and square jaws with prominent supraorbital ridges. The female skull has relatively round jaws with round orbits compared to the males (Fig. 2.7).

5. *Impact*: The skull is an extremely rigid structure capable of withstanding good amount of impact but has a limit. Hence load carrying to the tune of 15–20 k is relatively safe, but for impact there is every possibility of the skull cracking and causing damage. This finds pertinence in protective head gear design.

## 2.12  Adorations for the Face

Apart from all the factors mentioned for head the following ergonomic issues for face adoration play an important role in design (Fig. 2.8).

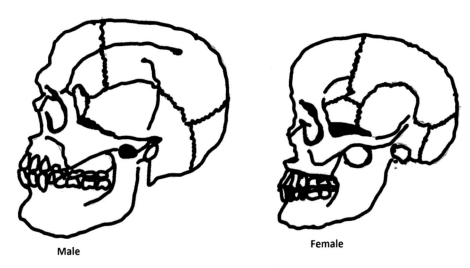

**Male**                                          **Female**

**Fig. 2.7**  Structural differences between the male and the female skull

**Fig. 2.8**  Ergonomics in facial adoration design. *Source* Photo by Malik Skydsgaard https://unsplash.com/photos/gC985WK2Rqo

1. *Gesture*: The most primitive mode of human communication is gesture, and in these facial expressions play a major role. So, adorations for ear, nose and forehead should facilitate such gestures and not stop them. Else users would feel uncomfortable in communicating in public domain. Design of adorations of this part should not be such that small facial expressions look amplified to the public. Thus, such adorations should "fit" the profile of individual face, and parts of the face and on the surface of the face and forehead should not protrude out.

2. *Weight*: Too heavy adorations (no scientific study available till date which indicates the weight-bearing capacity of the nose and ear lobes when pierced) for the nose and ear lobe should be avoided as it could lead to injuries. This could result from the weight of the product, vigorous movement of the body or movement in crowded places.

3. *Thermoregulation*: Due to the presence of large number of sweat glands in the face area, including the forehead, it is essential that adorations should not cover more than 2–5% of the total area of the face. This might impair sweat evaporation and lead to heat stress in a hot and humid environment.

4. *Body posture*: Movement of the torso and different body parts when vigorous leads to change in facial expressions. This happens in gyms where users work hard for training themselves. Thus, adorations should be able to account for these activities. For example, if a lady goes to the gym to shed extra weight, then she would be working very hard on the abdominal exercises. To do those heavy adorations might come in the way of the movement and create a nuisance for the lady.

## 2.13  Neck Adorations

Neck is the connecting link between the torso and the head. It is again a very dynamic part of the body which facilitates rotation of the head. As a tradition majority of the users prefer neck adorations in the form of necklace, pendant, locket, and so on. Nowadays women hang gadgets like pen drive, MP3 players, etc., from their neck. The following ergonomic issues need to be considered while designing such adorations, apart from what we have discussed in other chapters (Fig. 2.9).

1. *Load on the neck*: The neck is essentially a portion of the vertebral column support the body and having the trachea (breathing apparatus) and food pipe in front. As it is a delicate structure the load on the neck should be minimal as possible, else chances of injury or discomfort are maximum.

2. *Profile of the neck*: In growing or adolescent children and especially males a small protrusion in front is noticed which is called the "Adams apple". This protrusion needs to be factored in case adoration is designed to cover the complete neck portion in case of males. So, in designing such products a gap must be left at the Adams apple or the design should fit the curvature of the part of the neck. Extreme caution must be exerted

**Fig. 2.9** Ergonomics of neck adorations. *Source* Photo by Max Ducourneau on Unsplash woman wearing gold necklace and white shirt photo—Free Jewelry Image on Unsplash

in case of children as to the fact that they are not throttled while fastening such product around the neck. As a precautionary measure it should be loosely tied around the neck.

3. *Movement at the neck*: Neck is that part where lots of dynamicity take place and the degree of which varies. For some users forward tilt of the head leads to three folds at the skin at the neck, and for others such movement leads to more or lesser number of folds. Hence products should be designed for individual so that such folds do not come in the way of movement of the head.

## 2.14  Adoration for the Torso

The torso or upper part of the human body comprises the entire thorax and abdomen and the limbs including the shoulder. It houses major organs of the human body, like the heart, lungs, liver, alimentary canal, gall bladder, pancreas, kidneys, etc. This is the part after the face which adds value to the body in terms of beauty. In case of females, it is the breast and for males the chest muscle and the trim waist. The following elements of the torso need to be considered in the design process: (Fig. 2.10)

**Fig. 2.10** Ergonomics of torso adoration. *Source* Photo by BARBARA RIBEIRO from Pexels https://www.pexels.com/photo/sensual-woman-in-stylish-swimsuit-against-palm-trees-6485240/

1. *The breast/chest*: In women supporting the breast becomes very important for the fashion designer as it is a protrusion from the body in the forward direction. At times heavy breasts lead to sagging and often make the person stoop forward. So, any adoration of this part must confirm to the anthropometric dimension of the breasts, else it might look ugly. Beneath the breast lie the vital organs like the heart and the lungs. In case of males the chest is provided with muscles. In both males and females, any adoration of the chest is important as this is one of the most prominent parts of the body having the maximum surface area. For males this is also similar except that this part is muscular. So, adorations of this part are most noticed. This part also needs to be protected from cold environment else chances of getting a chest infection are high. This being another highly dynamic part needs to be dealt with accordingly. So, dynamicity must be added to the garment to allow for breathing, and for sports garments the breathing is far rapid and hence requires more space. In winter the chest area should be covered as this is the area which loose heat fast from the body, and hence it needs to be insulated.

2. *The hand-arm system*: This part connected to the torso has three joints at the shoulder, elbow, and wrist and hence is dynamic. Focus must be given to these parts because of dynamic movement. Lot of sweating happens at the arm pits and the elbow joints; hence it is essential to keep a way to ventilate these areas. For example, if the arm pit area of a garment is provided with perforated material, then it would dissipate lot of heat and keep the body cool. Similarly, the inner side of the elbow joint can be made of a relatively thinner material to facilitate ventilation with thick material on the outer side to prevent against wear and tear. This part of the body also undergoes rotational movement, and hence the garment should facilitate that. Wrist is another area which is dynamic and sweats a lot in summer. Hence watches, bangles, etc., in this area should preferably be perforated to facilitate ventilation. This area being relatively sensitive should be taken care of while using metallic products which are good conductor of heat and cold. The user might feel extreme hot or cold, in that case.

3. *Wrist*: This is another area where there are many sweat glands. Hence wrist watches or cufflink of full sleeve shirts gets dirty the maximum at this joint. It is better to keep an avenue for ventilation through porous design of elements at this part. If your watch is too tight over the wrist, movement of wrist is impaired; if tool is loose, it slides back toward the forearm or dangles near the wrist.

4. *Palm*: Adoration of the palms in the form of gloves both in the summers and winters is very common for all. Women are mostly fond of this type of adoration. The human palms are full of receptors, and the most sensitive of these receptors (which receives information from the outside world when the palm touches any object) are located at the fingertips and in between the fingers. Hence if palm adorations are thick, then the receptors are covered and cannot function properly. This leads to less sensation on the palm, and thus the grip on any object is reduced. You will notice this among women riding scooters in summers and wearing long gloves have a tough time in applying the clutch and brakes. Thick gloves also lead to reduced flexibility of the palms, and hence the palm cannot completely control any objects that she/he holds in the palm. Hence while designing, one should take care of this attribute. One possible ergonomic solution could be to keep the thickness of the glove's constant at all places, but the places in between the fingers could be made thin. This space normally does not encounter any surface, but at the same time facilitates coordination between the receptors located in the space between the fingers and hence increases the flexibility of the palm. Palm is another part which requires ventilation in hot environment. If it is entirely covered, heat cannot escape, and this makes the palms uncomfortable. In this case openings/perforations could be made on the outer side of the palms (dorsal side) through which the hot air could escape, and fresh air could also enter.

## 2.15 Adoration for the Lower Part

The lower part of the body comprises the hip bones which are a prominent anatomical landmark in case of females, as they protrude out from the body the maximum. In case of males the hip bones are much more toward the midline of the body, then comes the genitals, and then the two legs descend. The legs comprise thigh bones called femur, then the knee joint, then the two bones tibia and fibula which finally converge to the ankle, and this then extends to form the feet and the fingers. There are ergonomic considerations for designing adorations for this part. We will talk about general principles, and these will be discussed in depth at the relevant places (Fig. 2.11).

1. *The hip and the waist*: This part is critical in women as the shape of this part dictates how sexy the women would look like in public domain. This is followed by the buttocks which again are critical in both males and females, from the viewpoint of attractiveness. Imagine a man with flat buttocks, how ugly he looks. A woman with flat buttocks looks equally ugly. The reason is that the trochanter bone defines the symmetry of the human body into parts, the upper and the lower parts. For some people the upper part is longer than the lower part, and for some the lower part is disproportionately longer than the upper part. In these two cases for the body to look proportionate there could be two ergonomic interventions. The first one for the males could

**Fig. 2.11** Ergonomics in lower part adoration. *Source* Photo by Sonuj Giri on Unsplash silver-colored bracelet photo—Free Foot Image on Unsplash

be that he leaves the shirt fall over the trouser instead of tugging it inside the trouser. For females wearing of long garments extending till the knee would reduce this anomaly. This is also applicable for the males as well.

2. *Upper and lower legs*: These parts have two joints: the hip and the knee joints which are dynamic. Hence dynamicity must be provided for movement of the hip and knee joints and especially for the knee joint which is very dynamic during movement. The lower part of the body also requires a good degree of ventilation, and hence in a tropical climate the adorations should be such that they facilitate entry of cool air from below.

3. *Ankle and the feet*: These are dynamic as well and in fact support the maximum weight of the body. A good amount of heat is lost from the feet as well, and hence in a tropical climate it is necessary that the feet are well ventilated. Overheating of the feet can lead to heat-related disorders in the body.

## 2.16   Ergonomic Principles

(a)   Body dimensions are very important in fashion design.

(b)   Consider the point of contact to determine the relevant anthropometric dimensions.

(c)   Human body is heterogenous, thus take dimensions of all relevant body parts.

(d)   Do not make the mistake of taking all dimensions from 50th percentile.

(e)   User male and female data separate for calculation of dimensions of products.

(f)   Factor in dynamicity of different body parts.

(g)   Think of the different context of usage and the dynamic anthropometric dimensions.

(h)   Factor in the physiology of the body associated with thermoregulation.

(i)   Adorations should be safe and comfortable for users.

## 2.17   Assignments

Select a trouser and list down the pertinent anthropometric dimensions after a thorough task analysis with three users. Resize the garment and optimize it for production. You are to design this garment for both male and female use at the same time (wife wearing husbands' trousers). The waist size of the trouser is 95 cm and that of the husband is 94 cm and that of the wife is 98 cm.

   *Design Directions*:

(a)   Note the point of contact while wearing this product.

(b)   Consider male and female dimensions separately.

(c)   Keep in mind the principles of reach and clearance.

(d)   Optimize the dimensions keeping in mind the extra allowance needed.

(e) Design a crown for a fashion show from an ergonomic perspective keeping in mind the issues of comfort, safety, stability, and aesthetics.

(f) Select any adoration for the face like face scarf or jewelry and identify the ergonomic issues in them. Suggest ergonomic solution for the un-ergonomic issues.

(g) Select specific adorations for the neck, torso, and lower part of the body. What ergonomic improvements can be made to these parts to further increase their functionality?

## Bibliography

Gupta D (2014) Anthropometry and the design and production of apparel: an overview. Anthropometry Apparel Sizing Design, 34–66

Koca E, Kaya Ö (2016) A research on ergonomic approaches of apparel designers. In: Advances in ergonomics in design. Springer, Cham, pp 323–335

McGhee DE, Steele JR, Zealey WJ, Takacs GJ (2013) Bra–breast forces generated in women with large breasts while standing and during treadmill running: implications for sports bra design. Appl Ergon 44(1):112–118

# Ergonomics in Fashion and Accessory Products: Comfort and Functionality

**3**

**Overview**

This chapter discusses the physiology of thermoregulation and its pertinence in apparel design. The focus goes on to different aspects of thermal regulation which is often neglected by the designers. Concepts of homologous modeling are also discussed here. Ergonomics can be applied in different domains of the fashion industry, and this is discussed in this chapter here. The ergonomic principles are different, and how they can be applied with prudence makes the garment successful for the sector.

## 3.1 Heat Exchange

Heat exchange between the human body and the environment is very vital to keep the body at equilibrium. Extreme hot and cold are hazardous for the human, and hence to keep the human body in thermal equilibrium, it is necessary to lose and gain heat as and when necessary. The heat exchange in the body is dependent on several factors like the amount and type of clothing worn, fat content in the body, type of food consumed, and amount and type of physical activity performed. It is also affected by the ambient temperature, relative humidity, and air speed.

The human body loses/gains heat from the environment by three different methods, namely conduction, convection, and radiation. Conduction is the method of heat exchange when the body meets any object. If the object is colder than the body, then the body loses heat to the object. If the object is warmer than the body, then the body gains heat from the object. Convection is the method of heat exchange in which the air around the body affects the heat exchange. If the air around is hotter than the body, then the body gains heat, and if cooler, the body loses heat. Radiation is a method of heat exchange from a heat source like the sun, or fire, in the same manner as discussed above.

Hence the main question here is that when a garment is designed for the body, then it should not only adore the body, but also protect it from different environmental hazards, especially heat. In a hot environment the body tends to lose heat through the above process as mentioned and by secretion of sweat. Thus, these two activities should be facilitated and not blocked through design. For example, if you are a daily passenger traveling in India in the month of May in a bus, then your clothes should be loose fitting to permit heat loss from the body as to facilitate proper blood circulation so that the sweat glands are active and produce sweat. The sweat glands require proper blood flow to remain active. Similarly, if you are traveling in extreme cold, then you need to cover your extremities especially the head, feet, and the palms which loose lot of heat. In this condition tight fitting clothes like jeans trousers would not cause any harm. Thus, the purpose of loose clothing like shirt not tugged inside the trouser facilitates exit of body heat which is lighter compared to the ambient air from near the neck region and at the same time facilitates the entry of cool heavy air from below the untagged shirt, thus keeping the body cool. In winter tugging helps by preventing this convective flow of air in the body (Fig. 3.1).

## 3.2  Clo Value

Clo value expresses the clothing insulation of a given fabric used for making garments.

Clothing and clo values are indicative of thermal comfort of a person. It has been found that porosity of fabrics, i.e., the water vapor transfer through clothing, affects its insulation value as well. Even thickness of the fabric affects its insulation value. It has also been

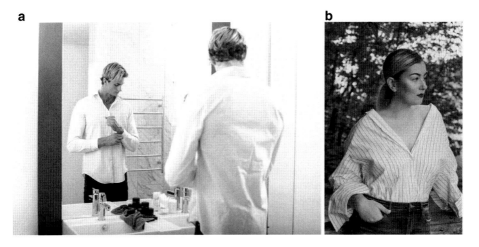

**Fig. 3.1** Convective flow of air in tugged and untugged shirt. *Sources* **a** Photo by Lumin on Unsplash man wearing white dress shirt photo—Free Shirt Image on Unsplash; **b** Photo by Valerie Elash on Unspalsh https://unsplash.com/photos/pFPvIlth4ecxt

observed that if the fabric insulation is high, then heat gained by the body by direct radiation is also reduced. Different garments have different clo values, and these are available in any standard textbook on the subject.

## 3.3   Effect of Clothing on Heat Exchange in the Body

The clothes that we wear affect heat exchange in the body. Insulation in the clothes reduces heat gain from the external environment and heat loss from the body. It has been found that the fiber type and design determine the degree of insulation to the clothes. A good design of fibers traps a good amount of air making heat exchange difficult. In fact, the insulation value of clothes is proportional to the linear function of its thickness. When fabrics are made of compact material or are soaked in water/wet, it clears out the trapped air thus rendering the fabric a bad insulator. In hot and humid environment, the garment should facilitate evaporation of sweat. Garments for cold environment should not become wet, else it will render the garment a very bad insulator and heat will dissipate from the body.

## 3.4   Thermal Comfort and Sensations

When it comes to comfort with a particular garment, there are individual differences, which need to be factored in. This level of comfort is dependent on many factors like type of work being done, food taken, environment, undergarments and existing clothing worn, season of the year, and acclimatization to the given environment. So, the designer must keep in mind these issues while designing garments. For example, the preferred room temperature in Singapore is higher than that in the USA or the UK. So, if you are designing a common winter wear for Singapore (let us assume!!!) and the USA and design it with a little bias for the people of Singapore, then the American people might feel a little warm in the winter garment.

Draught is unwanted cooling of the body caused by air movement. This at times causes discomfort, and hence garment design must take this into account. Human body is most sensitive to draught/air movement at the head region (including neck, shoulder, and upper back). It has been seen those subjects with short hair/less hair (to protect the neck from cooling) were less sensitive to draught than those with long hairs. Thus, the designer should not only focus on garments for giving comfort to their customers but also need to focus upon designing protective devices for the body extremities like head, feet, hands as well as these are the avenues through which heat exchange takes place. Thus, only apparel design alone will not provide comfort to the human being, but other aspects of the human body must be considered along with them.

## 3.5   Homologous Model

Human body dimensions are used in different facets of apparel and lifestyle accessory design. It is essential that to fine-tune the design in tandem with the varied dimensions of the user group, data such as height, leg length, hip/chest circumference, etc., are factored in. So far considerations were only made for different body dimensions in two dimensions. Now the time has come to factor in other variations in the human body where one is fat, other is thin, one has a broad shoulder, and the other has a very narrow shoulder. This is where the concept of the homologous model comes to the picture wherein mathematical data are used to model different parts of the human body and come up with accurate design of different fashion and lifestyle accessory products which can be customized for every individual.

## 3.6   Application of Ergonomics in Different Domain of Fashion

The principles of ergonomics can be applied in different domains of fashion to enhance functionality and ensure seamless integration of fashion and functionality with the users.

### 3.6.1   Medical

Ergonomics plays a very big role in hospital and medical domain. The application of ergonomics can be broadly classified into the following areas: work wear for doctors and supporting staffs, work wear for patients, and additional wears like gloves, aprons, facemask, and head gears for the operating staffs. Application of ergonomics has a very critical role to play in ensuring a sterile environment and keep the patient healthy. The major function of clothing is to maintain a static air layer next to the skin. This affects the insulation value of the clothing ensemble which might then increase or decrease, depending upon the amount of air trapped between the body and the garments, between fabric layers in a single garment, or between layers of different garments.

### 3.6.2   Color of surgeon's Dress

The color of the surgeon's dress, especially the gown and gloves are not white but either green or blue nowadays. Have you wondered why this is so? This is mainly because blood and similar body fluids in the presence of bright light and in the background of white gown look too bright and distracting to the human eye. This could lead to after images for the surgeons and the supporting staffs leading to further surgical error. Thus, shades of blue and green are normally used in surgical dresses (Fig. 3.2).

**Fig. 3.2** Ergonomic issues in the surgical dress. *Source* Photo by Olga Guryanova on Unsplash two men wearing blue lab coats photo—Free Medical Image on Unsplash

### 3.6.3  Surgical Dress

The main purpose of the surgical dress is to keep the patient sterile from the microbes and bacteria in the body of the surgeon and vice versa. At the same time, it should facilitate proper ventilation and heat exchange in the body of the surgeon and the patient. This will help in minimizing stress (cognitive) on part of the surgeon while performing any operative procedures. Thus, the quality of the fabric as well as loose-fitting work wear is necessary for facilitating heat exchange between the surgeon's body and the environment.

### 3.6.4  Head Gear and Facemask

The head and face gears are meant for not allowing hairs and sputum of the surgeons from falling into the open wound of the patient and thus contaminating it. At the same time as maximum heat is lost from the head, the head gear should be designed to ensure maximum ventilation and hence should be made of a fabric which facilitates that. The face mask should be loose enough by considering the dynamic anthropometric dimension of the jaws so that the surgical team might talk to one another when required during the operation.

### 3.6.5  Work Wear for the Healthcare Professionals

Apparel plays a very important role in the hospital and especially among the healthcare providers. Apart from protecting the providers against different pathogens from the patient it also protects the patients from any cross-contamination from the healthcare providers. So, when it comes to apparel in medicine that of the surgeon is the most critical. Apparel design for this group should follow the same principles of apparel ergonomics as has been described before. In fact, in all the domain of medical science the same principles of apparel design can be followed. It requires some special approach when it comes to designing the apparel for the surgeon in the operation theater.

There is a huge impact of the medical clothing on the operating performance of the surgeon, and it can mean two extreme conditions of life and death of a patient. For a successful surgical procedure there are many human factors attributes that are necessary. We are not going into those other factors. If we focus on those other factors related to apparel ergonomics, then mainly three factors dominate: the sterile environment of the theater, patient comfort, and comfort of the surgeon and the surgical team. The surgeon must work at different odd postures, and at times the duration is long lasting for hours in complicated surgeries. At such odd postures there is a rise in the metabolism in the body and hence more emission of heat. If this heat is not allowed to dissipate and cool the body, the surgeon might be under heat stress, and it could lead to surgical error. So, in essence the thermal environment inside the theater must be comfortable for both the surgeon and the patient. The patient lies in an almost seminude state. If the temperature of the ambience is too low, say below 21 °C, then it could be fatal for the patient. So, the only solution to this problem of a comfortable thermal environment is to keep the ambient temperature at a level where the patient is comfortable (not too cold) and play around with the apparel design of the surgeon. Currently the apparel (gown) of the surgeon is meant to be sterile so that entry of pathogen is not possible. This clothing must be impermeable so that body fluids from the patient do not come in direct contact of the surgeon and infect him. Such impermeable clothing leads to accumulation of heat in the body leading to discomfort, heat stress, and ultimately leading to medical errors and patients' death. Unfortunately for the surgeon the duration of operation most of the time extends to hours leading to increased discomfort and heat stress.

At the time of selecting the textile for surgeons, it is important to consider the fabric in such a way so that it can protect against pathogens (both ways), acts as a barrier between the task and the patient, and should ensure thermal comfort for both and especially for the surgeon as he is subjected to stress because of the reasons mentioned above. Such thermal stress on the surgeon can be reduced by introducing cooling system in the surgeon's gown through provision of a network of cool air or dry ice in pouches to keep the body cool.

In other areas similar principles of ergonomics can be applied ranging from the dress of the nurse, attendants, doctors on round and other paramedical staffs on duty. These principles in terms of biomechanics and anthropometry have already been discussed before.

## 3.7  Sports

Apparel ergonomics in different sports activities acts as a layer of protection and should ensure that at the same time it does not reduce the speed and accuracy of the user. There are different sports activities, and apparel ergonomics in all plays a crucial role. Here we are only going to discuss the ergonomic issues in sports apparel, and the different sports gear and their ergonomic design issues will be discussed else-were.

### 3.7.1  Sports Bra

During sports the mass and the movement of the female breasts play a very important role. These movements of the breast as the body moves need to be controlled, else it could be a problem for the sportsperson. So it is essential that one bra would not fit for all different types of sports activities. It would be guided by the type of sports and the anthropometric dimensions of the body specially the breasts (Fig. 3.3).

**Fig. 3.3** Ergonomics of bra design. *Source* TextPhoto by Womanizer Toys on Unsplash woman in black bikini holding bouquet of flowers photo— Free Underwear Image on Unsplash

The design of the bra should be in tandem with the breast dimensions keeping in mind that it keeps them stable but at the same time ensures sufficient space for blood circulation to take place, else it can lead to pain and discomfort to the user. In fact the ventilation aspects should also be factored in while designing bras for tropical climates and for sports like soccer or cricket which goes on for longer duration.

### 3.7.2  Sports Jersey

Designing of jerseys should also ensure loose fitting for facilitating proper ventilation and at the same time ensure that the body is protected from trauma during the game. The design should enable the players move their body parts as per the demands of the game.

### 3.7.3  Sports Pant

The ergonomic issues related to designing of pants include good fit based on anthropometry and the degree of dynamicity dependent on the type of task to be performed. Hence sports pants meant for jogging or other activities should be loose in fitting and preferably be open below in a tropical climate to facilitate entry of cool air. Slits inside the pockets would permit exit of hot air from the groin area, while the user is involved in rigorous activities.

## 3.8  Clothing for the Children

As children grow very rapidly a dynamic clothing design approach is required through the help of ergonomics. This growth does not take place in stature but also in individual body members and involves growth in circumference and width of many of the body members. Thus, such clothes should focus upon the usage of stretchable fabrics or the usage of loose-fitting clothes which could accommodate the growing body parts. This approach will be successful to an extent till the child attains adolescent. After this there is a spurt in growth which demands designing clothes as per the user's anthropometric dimensions and developing size charts for the users. There is certain work wear like socks, hand gloves, head gears, and face mask which can be worn by a larger spectrum of the population when compared to shirts and trousers which demands specific dimensions of the target users.

## 3.9  Ergonomic Principles

(a)  Heat exchange through the clothing is important in maintaining body homeostasis.
(b)  The clo value of fabrics plays an important role in thermal comfort.

(c) Homologous modeling can be applied to ensure better fit of clothing ensembles.

(d) Comfort and functionality of clothing go together.

(e) Anthropometric dimensions play crucial role in ergonomic design of different work wear.

(f) Dynamicity needs to be factored in at the design stage.

(g) The context of use of the product dictates additional ergonomic features.

## 3.10  Assignment

1. Pick up any fabric and try analyzing its thermal properties. How do you check for the thermal properties of an imported fabric in local condition? Apply the principles of homologous modeling in designing garments for the armed forces. Discuss in detail the different steps.

   (a) First observe.

   (b) Analyze how this would be used by the target users in actual context.

   (c) Check for thermoregulation of the garments.

   (d) Apply the principles of homologous modeling.

2. Do an ergonomic analysis of an apron used by the medical laboratory technician in the pathological laboratory. Identify the ergonomic drawbacks and its design and suggest suitable ergonomic design solutions for the same.

3. Select any sportswear and do an ergonomic analysis of the same.

4. Incorporate ergonomic design issues in a trouser to be used by an elderly at home.

5. Use ergonomic principles in designing a short for boys and girls to be used interchangeably.

## Bibliography

Kouchi M, Mochimaru M (2011) Errors in landmarking and the evaluation of the accuracy of traditional and 3D anthropometry. Appl Ergon 42(3):518–527

Motti VG, Caine K (2014) Human factors considerations in the design of wearable devices. In: Proceedings of the human factors and ergonomics society annual meeting, Vol. 58, No. 1. Sage CA, Los Angeles, pp 1820–1824

Wang Y, Wu D, Zhao M, Li J (2014) Evaluation on an ergonomic design of functional clothing for wheelchair users. Appl Ergon 45(3):550–555

# Ergonomics of Sensory Motor Functions in Fashion for the General Population, Challenged, and the Elderly

**4**

**Overview**
This chapter introduces the readers to the different application of visual ergonomic principles like visual cone, scanning and detection of the eye, color blindness in the domain of fashion, and with specific emphasis on the ergonomics of fashion products in the showroom. Fashion for the elderly and the physical challenged is at times neglected areas in the society today. Ergonomics as an applied science has a lot to offer. This chapter gives an overview of different ergonomic principles to be kept in mind while designing clothing and other lifestyle and accessory products.

## 4.1 Introduction

In the field of fashion design, the first look is the most important which appeals the maximum to the users. It could be going for a shopping to a store or a shopping mall, wearing a new set of lifestyle and accessory products, or attending a special occasion with a specific set of clothing and adorations for different parts of the body. In all these cases the first look and feel and the impression that it creates are very important. Visual ergonomics is that branch of ergonomics which deals with different ergonomic issues of the human eyes like visual cone, scanning and detection, pattern recognition, etc., which impacts this look and feel of the products. Thus, application of visual ergonomics is very crucial in determining that different fashion products can appeal to the users in a manner they are supposed to be or in a manner that the designers want them to be appreciated.

© The Author(s), under exclusive license to Springer Nature Singapore Pte Ltd. 2023
P. Mukhopadhyay, *Ergonomics in Fashion Design*, Design Science and Innovation,
https://doi.org/10.1007/978-981-19-4534-2_4

## 4.2  The Visual Ergonomics Route

The real journey of a fashion product starts when the product leaves the manufacturing hub and ends up in a showroom, or a shopping mall and is made visible to the target users. Thus, among a plethora of products the specific product has to "pop" up from the surrounding as if saying "I am here". The next stage is that the user picks up the product, tries on own self by looking the mirror, and tries to imagine how it would fit her/him if it was meant for a particular occasion, like marriage, office party, etc. After this the product, if finally selected, is taken by the user home. So, then the user is at liberty to use it as per his/her choice. When it is finally worn by the user, then the product is displayed to the outside world, comprising different types of users, a different variety of ambience, etc. After repeated usage of the product the same losses it lusters and shines when the user decides to dispose it off and opt for a new product. Thus, the product "retires".

## 4.3  Visual Cone

For any product (including fashion product) to be visible it should be within the visual cone of the users. Thus, if products are displayed in showrooms or shopping malls on racks, windows, etc., you need to ensure that they are within the visual cone (Fig. 4.1). To calculate this visual cone, follow the following steps:

(a) Decide where from (mark on the floor) you want the users to see the product.
(b) From that point take the eye height of users of different percentile (both male and female).
(c) At the eye height for different percentiles (male and female) draw a horizontal line parallel to the ground.
(d) On each horizontal line with a protractor draw a 35° angle above and 45-degree angle below the horizontal.
(e) Thus, you get the vertical cone of vision of each percentile.
(f) All the cones overlap each other, you must identify that common cone where products are visible to all.
(g) These must be done in stages with number of iterations.
(h) Similarly, the horizontal cone of vision is around 110°. So, after mapping the vertical cone map the horizontal cone from top view (plan view).
(i) Thus, you get the complete cone of vision within which your product should lie for the users to see.

**Fig. 4.1** Calculation of the vertical cone of vision.
*Source* Photo by Anna Shvets https://www.pexels.com/photo/three-women-standing-wearing-activewear-5012352/; https://www.pexels.com/photo/three-women-standing-wearing-activewear-5012352/

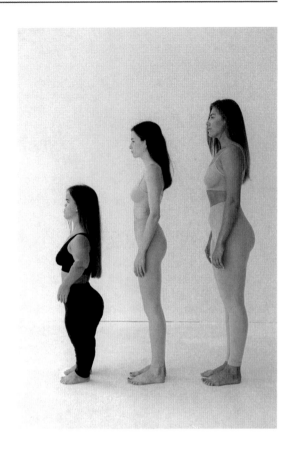

## 4.4   Scanning and Detection

The eye has a unique property of scanning any product from top to bottom. This is the reason why we prefer to first see the upper part of a person rather than the lower part. Thus, while displaying products, emphasis should be given to important information upper part compared to the lower part. While displaying adoration for the upper part of the body if there is lack of space, half bust mannequins can be used. For displaying accessories for the lower part of the body like footwear, jewelry for the feet, etc., the mannequin should be draped in a manner that the maximum attention is drawn to the lower part of the body. In such a seated mannequin on a mannequin on a platform at or near the eye height of the users would aid in a better scanning and detection (Fig. 4.2).

**Fig. 4.2** Placements of products for upper and lower part of the body with reference to the eye for ensuring better scanning and detection]. *Sources* **a** Photo by Andrea Piacquadio from Pexels https://www.pexels.com/photo/elegant-male-outfits-on-dummies-in-modern-boutique-3755706/; **b** Photo by Maria Orlova from Pexels https://www.pexels.com/photo/fashion-store-interior-with-garments-hanging-on-racks-4940756/

## 4.5   Visibility, Legibility, and Readability

All textual materials and other fashion and lifestyle products need to ensure that they follow these ergonomic principles. First, they should be within the cone of vision of the users as mentioned before. Secondly if text is used, then the individual letters should be identifiable as it is by the users lastly all textual materials should use proper spacing, background and foreground colors, etc., to ensure users are able to read them properly.

## 4.6   Ambient Illumination

To ensure that the product draws users' attention, the place of display should be well lit. If specific adorations are to be highlighted, then care should be taken to provide for task-specific lighting as well along with diffused illumination in the space.

## 4.7   Glare in Visual Field

Glare is the disturbance in the visual field. This can happen if luminaires are within the visual cone. Thus, care should be taken to ensure any light sources are away from the cone of vision. This could be achieved by calculating the vertical and horizontal cone of vision as mentioned before and then placing the light source beyond that. Glare can also happen when light falls on a shiny surface. This can be avoided to ensure shinny surfaces are replaced with textured surfaces.

## 4.8   Color-Blind People

A large portion of the population are color blind with red and green color blind believed to be maximum. Thus, in public places and when it comes to displaying of products or designing of fashion and accessory products pure red and pure green should be avoided as far as "practicable". The color-blind users see pure red and pure green in different shades of gray. Thus, one approach could be to make pure red and pure green a "little impure" by adding another color. This shifts the color spectrum from pure to something impure which is then visible to the color-blind users.

## 4.9   How It Looks in the Context

Users like to get a snapshot of the title picture as to how she/he would look in the product on her/his body. That is why mannequins are used to display different products. Full-size mirrors in the trial rooms also give the users feedback as to the exact look and feel with the product. To extend this further and to give the users feedback as to how it feels when the

**Fig. 4.3** Mirrors lined up to give a feedback to the users as to the look and feel of the product on their body when they move together]. *Source* Photo by Maria Orlova from Pexels https://www.pexels.com/photo/fashion-store-interior-with-garments-hanging-on-racks-4940756/

body is dynamic, series of full-size mirrors might be placed on either side, thus enables the users to walk up and down and if needed with their partners and to get an idea of how the new product looks on them (Fig. 4.3).

## 4.10  Introduction to Fashion Wear for Elderly and Challenged

Elderly and the physically challenged are very much dependent upon others when it comes to fashion and lifestyle products. Thus, the domain of fashion remains out of the reach of these spectra of the population. These users try to fit themselves not the lifestyle and accessory products including adorations which are mainly meant for the normal population. To do the users are not comfortable and at times are reluctant to go for fashion or stylish products. Application of ergonomic principles can go a long way in enhancing the usability of such products for the elderly and the physically challenged as well.

### 4.10.1   Dimensional Issues

Majority of the product available in the market is designed with reference to the anthropometric dimensions of normal population. Unfortunately, the anthropometric dimensions of the elderly and the challenged are quite different from the normal population. For example, as person ages the vertebral column bends forward. The joints become stiff, and it becomes difficult for them to wear clothes or adorations due to loss of flexibility. Similarly, for challenged people confined to wheelchairs or on crutches or walkers the movement of the different parts of the body is restricted, thus making it difficult to fit in to those products designed for normal users.

In these cases, the dynamicity of the body needs to be factored in. If users have stooped in the forward direction, then while designing shirts one should make the back part of the shirt a little longer than the frontal part, to ensure that while stooping the shirt does not go up and expose the inner wears. For those who are confined to the wheel chairs the dimension of the trouser should factor in the seating anthropometric dimensions and the associated movements thereof. The length of the trouser while seating should be taken and made a little longer so that they touch the feet as these users would not stand.

### 4.10.2   Force and Grip

Elderly and challenged people lose their power of grip and are not very good at precision grip demanding the dexterity of fingers. Thus, products designed for them like buttons, etc., should be avoided as they cannot operate them and better be replaced by Velcro. Similarly, earrings and nose rings should have magnetic fixtures which are easy to wear as they are unable to screw or unscrew with pinch or precision grip.

### 4.10.3   Visibility

It is important that while designing such products, the features should be obvious and visible to the target users. This is where product semantics comes into the picture in designing the products in such a way that its form tells the users how to use it with ease. If these features are made visible, then users would buy them and use them more, for example, the features of Velcro instead of buttons, magnetic strip in place of screw for earrings, magnetic rings for hair replacing the traditional hair clips, etc.

### 4.10.4   Maintenance

While introducing new features in the products, it should be ensured that their maintenance is not a problem, and even after repeated wear and tear the functionality and the features of the products are not lost. For example, if magnetic earrings are demagnetized, then it will fall apart. If Velcro strips used on shirts and trousers are damaged, then they make the entire products useless.

### 4.10.5  Ease in Wearing and Removing

For products meant for these groups of users it is important that they be worn and taken off with ease. You might have noticed how difficult it is for an elderly person to wear a trouser while standing as she/he is unable to maintain balance on one foot. If a user has one hand injured and must wear a shirt with one hand, then the opening of the shirt should be easy enough and involve no complex dexterous manipulation as such.

## 4.11  Ergonomic Principles

(a)  Products need to be within visual field of users.
(b)  Remember color-blind users.
(c)  Illumination should facilitate details of products.
(d)  No distraction in visual field.
(e)  Users like to see fashion products on them in context.
(f)  Keep things within reach.
(g)  Make thinks easy to operate with force exertion in one direction only.
(h)  No use of fine precision movement.
(i)  Easy visibility of the features.
(j)  Easy maintenance ensuring the special features are not damaged due to repeated use.
(k)  Factor in dynamic movement of the body.

## 4.12  Assignments

1.  A small showroom wants your expertise as a fashion ergonomist in advising them on how to display their new products for young college going boys and girls. These include shirts, trousers, churidars along with wrist watches, earrings, noise rings, etc.

    *Directions*:

    (a)  Decide from where exactly you want your users to see the products.
    (b)  Keep in mind color-blind users.
    (c)  Type of illumination.
    (d)  Height at which the products should be displayed.

2. Discuss the ergonomic features in the design of a fashion wear for a user confined to the wheelchair.

   *Directions*:

   (a) Take the relevant anthropometric dimensions while seated.
   (b) See the range of movement.
   (c) How would he wear and take of the product?
   (d) Does it cover all the parts of the body?
   (e) Does the wear look proportionate for the entire body?
   (f) Include features for some style statement like place for keeping hanky for easy access, medicine pocket in front, etc.

## Bibliography

Kouchi M, Mochimaru M (2011) Errors in landmarking and the evaluation of the accuracy of traditional and 3D anthropometry. Appl Ergon 42(3):518–527

Wang Y, Wu D, Zhao M, Li J (2014) Evaluation on an ergonomic design of functional clothing for wheelchair users. Appl Ergon 45(3):550–555

Tran TAD, Arnold M, Schacher L, Adolphe DC, Reys G (2015) Development of personal protection equipment for medical staff: case of dental surgeon. AUTEX Res J 15(4):280–287

# Fashion Ergonomics in Different Context

**5**

**Overview**

This chapter gives an overview of the application of ergonomics in fashion design for different context. It gives the readers an insight into different ergonomic issues in fashion design while travelling and exhibiting different types of movements. The chapter highlights the use of simple ergonomic design intervention in the clothing design for deterring against different crimes like pickpocketing, molestation, and rape against women is also discussed. The importance of user in ergonomic design intervention in fashion and accessory design is highlighted.

## 5.1  Introduction

Commuting to our workplace is a tricky affair especially with a new set or style of clothing or adorations. The users must run, board into a public transport, and reach their workplace in a properly groomed manner. Even when one must commute in personal transportation like car, two-wheelers, non-motorized transports like boat, and bullock cart, our clothing and adorations are all disturbed and need to be put in order. This dynamicity in travel to work needs to be factored in at the design stage, for the work wears and adorations. This is to ensure so that they suffer minimum distortions during the journey, and users enter their workplace with a renewed energy and vigor. The design of these products needs to factor in the different ergonomic issues to overcome some of the problems encountered by the commuters. We all know how embarrassing we feel entering the office if the knot of the tie shifts its position. We feel very bad when from the workplace, we directly go to attend a marriage party with all our clothing and adorations completely displaced, and we look devastated.

## 5.2  Crowded Transport or Space

We all need to step out of our homes for various purposes. It could range from going to our workplace, shopping, social gathering, movie, etc. Thus, we dress up accordingly at our home, in attire that we deem fit for the occasion. As a person, we want to be dressed properly in tandem with the occasion. That is why we check ourselves by standing in front of a mirror from different angles.

The context changes the moment we step out of our homes and move into public domain. In case you are taking a public transport it is natural to expect other people in the space. If the transport or the space is empty, there are no issues. In case it is crowded it maters how you stand within your personal territory without intruding into others, and the same is true with others. This ensures that your attire and accessories are intact. The problem emanates when the space becomes crowded, for obvious reason that it is a public place. With crowding and overcrowding the personal space around users starts shrinking, and ultimately a time comes when users intrude into each other's personal space and the body touches. This situation is also not static. Users must navigate through this crowd, which is like wading through water against a resistance. It is this resistance against the crowd which throws the neatly arranged attire out of gear, and it is disoriented, distorted, and loses its shape and form. So, when the user reaches his/her destination, she/he looks devastated because of the attire which is no more in its original form. Added to this there are issues of sweat, lost hairstyle, make-up erosion, etc.

The above context of a user is a common sight in every part of this world. Still fashion and lifestyle design must survive. One cannot give up saying that nothing can be done in such a context. Simple ergonomic design intervention can go long way in handling this situation if we use these ergonomic design directions at the nascent stage when the design of the clothing and lifestyle products is being conceived.

For the joints of the body, you need to factor in extra allowances for movements. For example, if you take the dimension of the sleeves of your full sleeve shirt just by letting the arms fall freely by the side of your body, you are probably mistaken. Consider the following context, which you have encountered at some point in your daily life:

(a) When you move in public domain, your hand moves, and even if your arm moves forward, your sleeves would roll back exposing a portion of your forearm(Fig. 5.1), which can embarrass you in public domain This is because you have not factored for dynamicity in movement of the arm. When the arm swings (as in a march past) because of the flexibility at the shoulder joints the overall length of the "shoulder-arm system" changes as it swings on either side of the free-falling arm. This is due to the increase and the decrease in the volume of the tissues at the shoulder joints. Therefore, you need to take this dynamic measurement while designing the apparel to ensure its proper fit to the body.

**Fig. 5.1** Ergonomic approaches in deciding the sleeve length of full sleeve shirts. *Source* Photo by Monstera https://www.pexels.com/photo/hands-of-black-people-in-black-and-white-shirts-6998462/

(b)  You take your seat in the bus. As you sit down, you find that the length of the trouser which was apparently touching the feet now rolls up and exposes your socks and "fortunately" stops just below the knee joint (Fig. 5.2). What an embarrassment! This happened because you failed to factor in the dynamicity which results when the knee joint bends. When the legs are straight at the knee (while standing), the length of fabric required is much less compared to when the legs are bent at the knees, because bending requires extra fabric, which if you do not provide, it would drag the trouser in the upward direction. Thus, while calculating trouser length, take measurements while standing, while walking, and while sitting down. If the person must squat at the workplace, make the person squat and take the anthropometric dimensions. This is because when limbs bend at the joints, their overall length increases due to elastic tissues at the joints.

(c)  A person has a big belly. If the person buys a ready-made shirt from the stores, there are chances that the shirt would not fit him well. In public domain, you would find him much more embarrassed when the tugged shirt comes out of the trouser when he tries to move fast or gets down the staircase in a hurry with dynamic movements. The reason for this is that the person with a big belly drags the front part of the shirt, thus

**Fig. 5.2** Trouser rolling back to the knees while seating. *Source* Photo by Atef Khaled https://www.pexels.com/photo/photo-of-man-squat-position-in-front-of-shopping-carts-1726458/

pulling it upwards (Fig. 5.3). If you look at the anthropometric dimensions, the person being an endomorph needs to wear a shirt with unequal length at the front and the back, with more length at the front compared to the back. This dynamicity needs to be factored in at the design stage. This extra length of the shirt in front would compensate for the extra fabric being pulled in the upward direction due to the belly, which pulls upon the fabric of the shirt. Thus, people with big belly need to wear shirts, which are long in front and a little short behind.

(d)  You keep your purse in your back pocket of your trouser. It is supposed to be the hip pocket, and the point of reference for the same should be the trochanter bone (Fig. 5.4). If you look at your old purse, which is being used, say for one year, you will notice that it has acquired a peculiar shape. Instead of being flat (which it was originally), it has buckled inside and taken a peculiar boat shape (Fig. 5.5). This is because the placement of the hip pocket in the trousers in most cases is the back with the gluteal furrow as the point of reference. Therefore, when you are sitting for a long time, your body weight falls on the purse, which in turn pushes into the seat padding. This in turn leads to inward buckling of the purse (Fig. 5.9). So ideally, you should observe the point of contact of the lower part of the body when person is seated and position at the pocket. The seat pan should not touch the pocket. Ideally, the pocket should be near the trochanter bone. This would ensure its easy access while siting and ensure that the weight of the body while seated does not fall on it.

**Fig. 5.3** Ergonomic approaches in designing shirts for obese users. Unequal length of shirt helps as the belly drags the shirt upwards. *Source* Photo by Ehimetalor Akhere Unuabona on Unsplash https://unsplash.com/photos/9LSTxzIaXiA

*Considerations for physiology*

As the user moves heat is produced in the body. In a tropical climate the sweat glands are activated leading to sweating which is a natural phenomenon for releasing the latent heat and cooling the body. The maximum concentration of sweat glands in the body is in the groin, underarms, cervical region (neck), palm, and sole of the feet. The feet are covered with footwear and socks which do not allow the excess heat to dissipate. The exposed parts look ugly with watermark becoming prominent as the person sweats. So ergonomic solution to this problem would be to provide sweat-absorbing layer of fabric like pure cotton and reinforce the areas that sweat maximum. This would absorb the sweat as it is produced and not let it reach the surface of the fabric. If necessary, these reinforced fabrics could be made detachable so that the user can replace it when it becomes too wet, on the concepts of using a sanitary pad.

## 5.3 Driving a Two-Wheeler

Careful observation of riding and traveling in a two-wheeler would reveal the following ergonomics issues:

**Fig. 5.4** Point of reference for hip pocket. *Source* Photo by Clear Cannabis on Unsplash https://unsplash.com/s/photos/purse-in-hip-pocket

**Fig. 5.5** Bending of purse because of wrong position of hip pocket at the buttocks and not at the hip. *Source* Photo by Matheus Ferrero on Unsplash https://unsplash.com/photos/TkrRvwxjb_8

(a) Legs remain on either side of the vehicle in a "saddle seating" posture. Due to this posture, the buttocks and the groin areas are in contact with the vehicle. Due to dynamicity in the ride the groin and the buttock area rub against the vehicle body. This can lead to stains in the trousers from the paint of the vehicle, and more than that this continuous friction can lead to thinning of the buttock and groin area and the trouser can split apart in these zones. Thus, the groin and the portion of the inner side of the thigh need to be reinforced with extra fabric from inside. This extra fabric acts as a cushion and protects against bruises to the tender skin in these two areas. Apart from this, it would also ensure that the fabric at these parts does not rip apart suddenly.

(b) The diameter of the trouser bottom can be trapped in the wheel of the vehicle if it is too large. This can often lead to serious accidents. Therefore, at the feet and vehicle interface there should not be any loose end of the fabric to be entrapped in the wheel of the vehicle. Thus, for such users the bottom of the trousers should be narrow to ensure that minimal fabric protrudes out.

(c) For women riders the hairs are an area of concern while wearing helmet. The hairstyle goes for a toss when the helmet is on the head. Thus, women riders at times are reluctant in wearing helmets. Even for male riders with stylish hair the helmet turns out to be a problem as it becomes very tight to wear such helmets. Ponytail helmets are common for women nowadays, which provides for a small opening at the back of the helmet for the ponytail (extra hair on women) to pass through. For providing some comfort to the users (both male and female) due to the thick layer of hair, some extra allowance needs to be provided over and above the anthropometric dimensions of the head, such as head circumference, head breadth, and vertex to chin length. The extra allowance to be added is dependent upon the thickness of the hair, but normally it can range from 2 to 4 cm over and above the anthropometric dimensions of the head as discussed in Chap. 2 (Fig. 5.6).

(d) In case when the women pillion driver is wearing skirt or similar clothing, then going for a saddle seating is not convenient. In that case, they sit at perpendicular to the rider. The ergonomic issue here is as the vehicle moves, the loose clothing for lower part of the body exposes the private parts of the users due to the gush of air, causing embarrassment on the road. The users need to continuously hold their clothing with one hand to keep the lower part covered. Ergonomic intervention here could include small Velcro strips attached on the bottom of the clothing, which could be brought close together and attached to the legs. When the user gets down from the vehicle, the leg movements would automatically open these adhesions between the Velcro pieces, and the clothing retains its normal position.

(e) For women users riding two-wheelers the time for menstruation is challenging in that they must wear an extra pad. Thus, design of inner wears and clothing should factor in extra allowance for the crotch height to the tune of 1–2 cm to ensure comfortable fit of the lower garments.

**Fig. 5.6** Anthropometric dimensions in the design of helmet

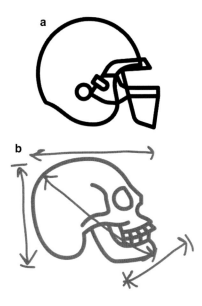

(f) For male users suffering from hydrocele (mild to severe) increases the size of the testes. Thus, the design of undergarments and trousers near the crotch and groin areas must factor this extra dimension so that when the testes meet the vehicle in saddle sitting, they can spread out uniformly in a thin layer and do not make the ride uncomfortable.

## 5.4  Riding a Car

Car driving and riding are a convenience and pleasure for many of us. It is because of the service it provides in navigating from one point to another either for pleasure, work, or any other purposes. Unfortunately, very few cars design accounts for different types of clothing worn by the users while driving or riding a car. It is here that the main problem emanates. Should we design the car, should we design the clothing, or should we design both? As fashion designers, we have very little scope in the current context to design the car interior from an ergonomic perceptive. However, fashion designers can incorporate some ergonomic features in the design of clothing to be used while driving or riding a car. The following are some ergonomic design directions for the same:

(a) Getting inside and out of the car is problematic while wearing loose clothing as it gets trapped in the car's controls like steering, gear shift, and at time in the seats and windows. Movement inside the car demands dynamic movement of the torso and the lower limbs. To address these problems, two ergonomic approaches could be there. You should ensure users wear clothing, which does not have loose ends, though this might not be feasible as it tends to impose upon the users a specific type of clothing. The second more viable ergonomic approach would be to ensure that in case loose clothing needs to be worn, then this loose portion of the clothing could be rolled and attached to the main clothing, and once the users step out of the car these could be opened. The attachments could be in the form of small Velcro strips or buttons, which can act as an accessory.

(b)   During entry and exit from cars, the body bends either forward or sideways. This elongates the personal space around the user. The body parts, which protrude out from this invisible personal space (if you look from the top), are the shoulder-arm system including the palm, buttock to knee length, and foot length. These are the anatomical landmarks which meet the protruding elements inside the car, namely the steering, indicator stick, gearshift, handles of the door, etc. Thus, at the design stage if it is ensured that at these protruding anatomical landmarks of the body, no accessories like pockets, big buttons, or other protruding lifestyle accessory products are not attached, then the chances of the clothing ensembles in getting trapped and causing movement restrictions are reduced.

(c)   Driving a car requires operation of the accelerator, brake, and clutch with the feet. This is solely based upon feedback through the feet. This feedback is reduced if users use footwear with very thick soles. Such reduction in feedback from the feet can create problems for novice drivers. Thus, it is recommended that novice drivers should use shoes with thinner soles while driving. In tropical climates drivers of public transport system use footwear with very thin sole, as they must operate the foot controls very often in a bumper-to-bumper traffic on crowded roads.

(d)   While riding a car, you might have noticed that the texture of the seats plays a very important role when you need to slide your buttocks on the seat to pave the way for other passengers. This sliding mechanism becomes easy if there is less friction between the fabric of your clothes and the fabric of the seat. Thus, glossy seat textures preferably in leather or Rexene over any other fabric as it reduces the friction between the clothing and the seat, thus making movement over the seat relatively easier. In taxis and other public transport system, this should be the preferred approach.

## 5.5   Rain

Wearing a raincoat over the normal clothing is a nightmare for many. Apart from the comfort, which goes for a toss, wearing it on normal clothing is a problem. The following are some ergonomic design directions to be kept in mind while designing clothing and coat for rains:

(a)   A raincoat is it be designed keeping in mind that it is to be worn above the existing clothing and hence should be easy to wear and remove. To ensure this, it is better that the coats dimensions are bigger than the anthropometric dimensions of the users, and they are loose fitting and provided with designing elements which makes it easy to wear. So instead of buttons which are difficult to operate with a wet fingers as it requires precision grips Velcro strips or zippers could be used for the coat in front.

(b) The sleeves of the coat should be loose as that would ensure its easy putting on and off and ensure that in a tropical climate, movement of the hands would facilitate exit of warm air from the body and cold air is sucked in through pumping action as the hand moves. In a cold climate, the loose sleeves might be closed with Velcro strips, near the wrist to retain body heat.

(c) Here it is better to make the raincoat loose by providing extra allowance with respect to the user's anthropometric dimensions. There could be additional slits near the armpits and the shoulder blade for facilitating dissipation of body heat and keeping the body cool. The raincoat should have minimum buttons and provided with Velcro strips so that wearing it is easy for the target users.

(d) The rain trousers are mainly used by those on two-wheelers or by the pedestrians. As they must be worn with the footwear, the diameter of the lower part of the trouser should be designed factoring in the length of the shoe. The lower part of the trouser could be kept loose in tropical climate to ensure movement of air due to pumping action or could be closed with Velcro strip near the ankle to prevent heat loss.

(e) Rain pants are to be worn while on the move. So to ensure easy wearing, the waist should be provided with elastic band that adjusts to different waist size and makes it easy to fit different somatotype of users.

## 5.6  Mobile, Wallet, and the Pick Pocketers

Users are now in different professions. Many professions demand a specific type of work wear, which need to be comfortable, and at the same time the style statement in them should not be lost. For all users their gadgets like cell phone, need to be kept in their clothing and they need to be safe as well. Users need to guard their valuables like purse from pickpockets in public domain. There are issues of crime against women in public domain. In all these cases, the apparel plays a very important role. Ergonomic design intervention in the form of small value additions can go a long way in increasing the usability of these apparels while keeping the fashion and style statement intact.

We all need to carry our wallet and mobile phones along with us, along with the keys of our house and office in key rings. For some these are carried in person, especially those who do not want to carry extra bags. For others it is carried inside bags. To prevent our valuable items from pickpockets in crowded places and crowded public transport, the following ergonomic design approaches could be investigated:

(a) Feedback is the key to ensure that our valuables are in place or have been displaced. Thus, clothing design should ensure that the user gets this feedback to his/her body. While designing pockets in the trousers and the shirts for keeping the valuables, it is

**Fig. 5.7** Inner pockets with
feedback to the users

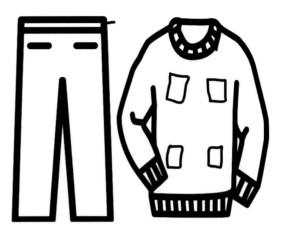

better if the pockets are designed on the inner side and in close contact with the body
(Fig. 5.7). This would have two advantages: Firstly it would not make it visible to the
criminals easily as it is not within their cone of vision, and secondly being in contact
with the inner garments they would provide constant feedback to the users about their
presence.

(b)  The same design for shirt and trouser pockets could be followed. Another preferred
     area for pockets in trousers for carrying valuables would be near the trochanter bone
     (attached to the inner lining of the trouser near the waist). Additionally, pockets could
     be positioned on the lower part say near the knee joints but toward the midline of the
     body (Fig. 5.8). This makes the pockets accessible to the users but difficult for
     criminals to access as it is not directly facing them.

(c)  As an option, one can have hip pockets but those could be provided with, zippers,
     which lock when, bend (Fig. 5.9). Additionally, if there are double pockets, that mean
     a pocket with two layers, you open the first layer and access the inside. You open
     another zipper, then access the inner most layer; here wallets and mobiles can be
     stored. The profile of the fingers and the palms is such that accessing a single layer
     pocket you need dexterity of the fingers. However, when you must move your palm
     and fingers through two layers in the pocket, that too blindly it becomes difficult and
     involves dynamic movement of the palm, which would alert the person.

## 5.7   Crime Against Women in Public and Private Places

Globally crime against women is on the mount. It is unfortunate to see that in crowded
places including public transportation systems women are the target for molestation and
other sexual assaults. In secluded places there are heinous crimes against women like rape
and molestation. The following ergonomic design intervention could be applied at the
design stage to protect women against such crimes:

**Fig. 5.8** Positioning of trouser pockets for preventing pickpocket (inner side of the belt and between the legs near the ankle

**Fig. 5.9** Double-layered pocket with zipper opening in two opposite directions

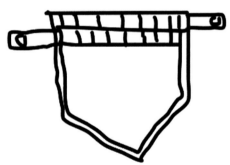

(a) Touching the private parts of women is an age-old crime common in crowded places, public transport, and at times at workplaces as well. However the law of the land is very strict when it comes to workplaces, where it is easy to nail the culprit, it is not so in public places. Thus, the ergonomic approach would be to deter crime in public places. Any offender normally targets the protruding parts of the female body, namely the breasts, buttocks, and to a smaller extend the genitals. There are two ergonomic approaches to this:

   1. In crowded places offenders mix in the crowd and touch these body parts at first by looking at them and then anticipating their exact position. This position is gauged by the stature of the victim. So, if clothing (special ones) is designed in a manner that these protruding parts are not visible easily to the victims, in that case such crimes could be deterred. One way to do this is to use certain patterns on the

**Fig. 5.10** Patterns to deter crime against women: molestation. *Source* Photo by Erik Eastman on Unsplash https://unsplash.com/photos/-6zFVL4YuaM

fabric, which leads to after images and creates illusion, and especially those that create movements. This the eyes cannot tolerate, and the culprits are unable to focus on specific body parts. These illusion creating images should be located at places like elbow, wrist, ankle, neck, etc., so that the focus from the protruding parts of the body shifts (Fig. 5.10).

2. Some specific colors could be used in clothing design (must be done at the fabric designing stage) like red and green/yellow (complementary colors) with which if the patterns of the clothing fabrics are designed, it would create after images. These images confuse the offenders as they lose their focus, and hence, they are unable to commit the crime in public domain, which demands focus.

For prevention of rape, lot of efforts are already in place. From the viewpoint of ergonomics, to deter rape, a layered clothing (Fig. 5.11) for the lower part of the body could be thought off. If access to the body is hindered through layered clothing, then the crime can be deterred largely. Criminals normally attack those women where it is easy for them to disrobe the women and then attack her. Two to three-layered tight undergarments also deter such crimes as it makes it difficult

**Fig. 5.11** Layered clothing for preventing crime against women

to remove layered clothing, which are tight by the fingers alone. It leads to fatigue of the fingers very early. Another ergonomic approach could be to ergonomically design a single undergarment comprising a panty attached to the bra (somewhat like a single piece swimsuit). This makes exposing of the lower part difficult for the criminal for committing the offense. The act requires pulling the undergarment down to expose the genitals. Making it a single unit and attaching with the upper part (bra) make it difficult to pull it down.

## 5.8  Security Personnel

Clothing for security personnel should factor in aspects of good fit, comfort, and convenience and act as a symbol of maintenance of law and order in the society. The following ergonomic design intervention should be integrated in the design of the clothing:

(a) To ensure proper fit, the clothing should be designed as per the anthropometric dimensions of the forces. They should be customized for everyone. Care should be taken to make some parts loose (add dynamicity) like the knee joint (trousers) and elbow joints (shirts) which must exhibit lot of movements while chasing criminals. The diameter of the trouser bottom should be narrow and less than half of the foot length so that while running the trouser does not get trapped in the shoes.

(b) There should be pockets at strategic locations for keeping different utilities like pen, whistle, notebooks, mobile phones, and other communication gadgets. As users are right and left handed both, there should be uniform distribution of pockets on either side of the midline of the body to ensure that the dominant hand is able to reach it with ease. For accessing the pens, the pockets could be designed on the shirtsleeves near the upper arm. This makes it easier to access in a much natural posture. For notebooks, the trouser hips are the best location. Additionally, the inner side of the tights could be used as well if confidential contents are written in the notebooks. This

**Fig. 5.12** Angular pockets for maintaining neutral wrist (shown with angular stripes in the illustration above)

area makes it accessible to the policeperson but makes in in accessible to unscrupulous elements who would like to access them. Whistles could be kept in the front pocket go he shirt at the level of sternal notch on either side of the chest.

(c) The trouser belt could be integrated with a small pouch for carrying a small drinking water bottle for those manning a busy traffic on the roads.

(d) Retroreflective strips can be attached to the joints of the body like wrist, elbow, knee, and ankle region of clothing to enhance the visibility of the personnel working at night. As these parts are dynamic and exhibit movements, they create pattern of humans from a distance and warn the motorists about the presence of people on duty on the road.

(e) Design of pockets for accessing pens or whistles could be made angular (Fig. 5.12), because in that case the access becomes faster and easier. The wrist need not deviate and thus maintain the neutral position while accessing the pens, whistles, or other belongings which requires frequent access.

## 5.9 Garbage Handlers

One of the most important parts of our society is those who handle the garbage that is produced every day. It might be the individual handler or those who come in groups in special vehicles. Here we are talking about those who handle domestic garbage. Their clothing or work wear as it is popularly known as demands some ergonomic design directions as well:

(a) The clothing should cover the entire body to protect it from the dirt around. The clothing should be loose near the waist and different joints of the body like, shoulder, elbow, knees, and ankles where in dynamic anthropometric dimensions need to be incorporated depending upon the range of movement of different parts of the body.

(b) The ends of the clothing should be secured with the body to ensure that no dirt enters the body. As the task requires lot of forward, upward, and sideward bending, the length of the clothing should factor in these dynamic movements.

(c)  The clothing ensemble should be one unit instead of having two different parts. This ensures less opening and gives better protection.
(d)  In a tropical country, users would sweat, and there need to be some air circulation in the body. If this is not taken care of, it leads to heat stress and loss in productivity. As a precautionary measure, two options are available. One is using vapor permeable fabric for such clothing, but these are very expensive. Another relatively low-cost option would be to use to provide zippers near the underarm, shoulder blade, neck, and the ankle. These zippers could be used to create opening in those part of the clothing, thus allowing the hot air to leave from above and facilitate the entry of cold air from the feet, thus facilitating air circulation in the body.
(e)  Retroreflective strips could be used as mentioned before to enhance visibility of such users when they work in the dark and near the main road.

## 5.10  User in Fashion Ergonomics

In the world of fashion, lifestyle, and accessory design, comfort is an important criterion. If users are not comfortable, how good looking the product might be, it would be rejected in the end. Thus, ergonomic principles lay emphasis on proper selection of skin-friendly materials, design intervention for heat dissipation, and usage of special vapor permeable fabrics work wear to ensure proper heat dissipation from the body. You need to keep in mind that the skin in the body is believed to be the protective layer, and clothing is believed to be the second skin. Thus, to maintain equilibrium in the body so that we are comfortable, body tends to lose and gain heat as and when required. When you design your products for the human body, you need to factor in the comfort aspects of the clothing as well.

When a fashion designer is to utilize these principles of ergonomics, he/she should first start with the systems ergonomics perspective of design. The user study is the key to any successful ergonomic intervention. One gets an idea what the users want, what the problems are. This should be followed by direct observation and activity analysis of the users to get an insight in the exact context of use. Task analysis gives an insight into the different postures assumed by the users and how much of dynamic anthropometric considerations the designer should go for. After this optimizing the anthropometric values for different parts of the body for the apparel is to be done. User trial forms integral part of ergonomics for getting users feedback regarding comfort, movement, and other ergonomic aspects. If one wishes to apply ergonomic principles in manufacturing, then the issues of workstation design, posture, and environment need to be applied. The designer must understand that each human body is different. This inherent human variability must be bridged if a design must fit a large spectrum of the population. Application of ergonomics teaches the designer to bridge that gap and account for this variability, thus equipping him/her with an additional tool, which must go parallel with the principles of fashion design.

**Fig. 5.13** Textured clothing for visually impaired. *Source* Photo by Isabela Kronemberger on Unsplash https://unsplash.com/photos/0rUc4_00L-A

## 5.11  Fashion Ergonomics for the Visually Impaired

Ergonomic principles are now used to bring fashion close to visually impair. Textures, embroidery, 3D printing, and similar technologies are used to evoke tactile responses for the visually impaired so that they may also enjoy different fashion and lifestyle accessory products. Ergonomic principles are being used to design special tags with brail writing for this group. Many clothing is having small bean like structures attached to the ready-made garments so that they act as a style statement and at the same time conveys specific information about the clothing to the target users (Fig. 5.13).

## 5.12  The Future

For fashion, lifestyle, and accessory design, ergonomic principles can help in bringing the products much closer to the target users. Ergonomics would help you to design your products from a user perspective. Ergonomics will help you to optimize the dimensions to a higher level of accuracy by factoring in anthropometric or human body dimensions. Ergonomic principles can be applied in almost any domain of fashion design with prudence and great benefits and making these products much more "humane" and bringing them closer to your target users. Ergonomics is a multidisciplinary subject, which helps you to enrich fashion, and this is achievable through bringing the user in the center of the design process. You can take off from here and apply ergonomic principles in the designing of footwear, headgears, and other accessories. If you apply the principles of ergonomics in your profession, then fashion and design do not remain only a style statement for the celebrities, rich, and stylish people as is normally believed to be. Fashion design then becomes an essential part of our lives, as it makes our life much more comfortable, useful, and happy.

One needs to understand that fashion and ergonomics should go hand in hand. Ergonomic principles should be applied from the very beginning of the design process and not in midway or at the end. One must understand that ergonomics as a subject makes your design much more enriched and full proof. It is up to you to take a call how much of ergonomics you would like to apply in your design. Last but not least, application of ergonomic principles in your design does not jack up the production cost as it is wrongly believed it be. On the contrary, ergonomics helps you to optimize your design and thus reduce the design and manufacturing costs largely.

## 5.13  Ergonomic Principles

(a) Consider dynamic anthropometric movements for designing for commuters in public transport system.
(b) Avoid protruding portions of clothing and adorations in public and crowded places.
(c) Fabric reinforcement to those places, which undergoes maximum movements.
(d) Provision for ventilation in raincoats.
(a) All clothing works in a context.
(b) Context decides body movement.
(c) Access for criminals to body parts to be reduced by covering the anatomical landmarks.
(d) Layered clothing gives protection against molestation and rapes.
(e) Visibility of the user from a distance is important in clothing design for specific professions.

## 5.14  Assignments

A. Your expertise has been sought for ergonomic design intervention in the clothing of a female executive who travel to office in a crowded train. She prefers wearing sari, salwar kurta, and at times shirt and trouser.

*Directions:*

(a) List down the touch points associated with each clothing.
(b) Try to link each touch points with relevant anthropometric dimensions.
(c) Optimize the dimensions with reference to different percentile.
(d) Add dynamicity based upon the movement.
(e) Ensure extra clothing materials are flushed with the main clothing ensemble.
    Use ergonomic principles in adding value to the clothes worn by the painters who paint house interiors.

*Directions:*

a. List down the tasks.
b. Identify the type of movements.
c. What products he needs to carry always.
d. Case to every product.
e. Which product is requiring frequently: Can it be attached to the clothing?
f. Allowance to clothing based on dynamic movement.

## Bibliography

Çivitci Ş (2004) An ergonomic garment design for elderly Turkish men. Appl Ergon 35(3):243–251

Singleton WT (ed) (1982) The body at work: biological ergonomics. Cambridge University Press

# Fashion Ergonomics Exercises with Guidelines

**6**

1. *Use ergonomic principles in designing a work wear for highway workers working at night, so that they can convey their presence to the motorists driving along the highway.*

**Guidelines**

People working at night under such conditions need to be provided with reflective dresses, the visibility of which is more at a distance. The norm is if workers are to be made visible during the day as well as night, then greenish yellow fluorescent jackets or work wear is the best. Now if the budget does not permit those, then fluorescent strips of fabrics can be used on the work wear in the form of a geometrical shape like triangle and rectangle as shapes along with fluorescent color acts as redundant information and improves detection from a distance. Humans are very good in recognizing patterns.

2. *Design a formal wear for women who travel to work daily in crowded bus and train and then reach her office. Use ergonomic principles to detail out your design.*

**Guidelines**

This type of problem should start with user's study. Do a complete observational study of how such women travel and what parts of their body are maximally active. Then do a detail task analysis to find out the range of movements the users exhibit. This would give you an insight into the requisite anthropometric dimensions to be considered and the degree of dynamicity to be considered. The next consideration would be for the environment. This would dictate the fabric type. Color of the fabric should match the stereotype for the profession and the work culture.

© The Author(s), under exclusive license to Springer Nature Singapore Pte Ltd. 2023
P. Mukhopadhyay, *Ergonomics in Fashion Design*, Design Science and Innovation,
https://doi.org/10.1007/978-981-19-4534-2_6

3. *Use ergonomic principles to design the pockets in any garment for holding mobile phone, pen, and wallet. Highlight the safety features in each against pickpocketing in a crowded public carrier.*

**Guidelines**

To address this problem, you need to do a detail task analysis of how different users use their pockets for storing different devices and how easily they can access them. To incorporate ergonomic design features in the pocket, you need to put yourself in the feet of the pickpockets. This would give some insights into their task analysis against which you can design some solutions. Pickpockets mainly use their fore and middle fingers to lift belongings from others pocket. The pockets could be covered with a flap, or it could have two layers of fabric overlapping each other so that entry of finger becomes difficult. From a dynamic anthropometric perspective, you might reduce unnecessary free space in the pocket so that movement of belongings becomes difficult. This again is to be done considering the need for your access for your belongings when required.

4. *Using ergonomic principles design apparel for women which would protect them from sexual harassment in crowded places.*

**Guidelines**

First identify the different organs of the female body which are vulnerable to attacks. For the upper part mainly the breasts and in the lower part the buttocks and the genitals are the targets. If through anthropometric dimensions you can reduce the exposure of the body for the breasts and cover it up a little higher, it could act as a resistance. The opening for the top could be in the form of a V rather than a semicircle as the former makes it difficult to insert one's palm and touch the breasts. Similarly for protecting the genitals the clothes could be made loose fitting which would make it difficult for the assaulters to guess the exact position of the genitals and the buttocks.

5. *Use ergonomic principles to design and develop a garment for the police force of your city. You must design the accessories as well.*

**Guidelines**

Security personnel are protector of the common people. The selection of color for these personnel should be guided by that principle. The color should command respect and hence should be in tandem with the mental model of the users of which color depicts to them the protectors. As to the functionality of the product a detail task analysis will reveal what is needed. The fittings of the clothes will be guided by the dynamic anthropometric dimensions as before. Task analysis along with postural analysis will reveal the position of pockets, their nature, and their dimensions. Other pockets or straps would be decided by similar task analysis.

The accessories for the gun, ammunitions, notebooks, and pen would be guided after a detail task analysis of the user under different circumstances.

6. *Design a work wear for the garbage handlers using ergonomic principles. The work wear should not only protect them but also give them an identity in the society.*

**Guidelines**

For designing such work wear similar principles as discussed before need to be followed. To create an identity, you need to go to different types of users to get an insight into the mental model with reference to color, fabric texture and any other design elements. The color of the wear should identify the users from the surroundings and be visible at night to the motorists driving down. It should permit easy ventilation and preserve body heat as and when required.

7. *Use ergonomic principles for designing clothing for the physically challenged and the elderly which they can wear independently.*

**Guidelines**

This should also start with a detail task analysis which would give an insight into the amount of dynamicity and the restrictive movement of some of the body parts. The strength of the user should be considered next which will dictate what type of materials should be used for closing and opening the garments, i.e., buttons, Velcro, push button, or hooks.

8. *Use ergonomic principles for designing clothing for the blind so that the other blind people can appreciate the clothing through its feel.*

**Guidelines**

To solve this assignment, you need to know that blind people communicate through their tactile (touch senses) of their hands. These receptors are at the fingertips, between the fingers, and all over the hand. So, the surface of the fabric should communicate to them what it is all about, woolen, cotton, jeans, corduroy, or anything else. In case of ready-made garments, the buttons (shape), number, and type would matter to them. All this information needs to be gathered after a detailed user's study on the blinds inside a shopping mall.

.

Printed in the United States
by Baker & Taylor Publisher Services